Fundamentals of
Aquatic Ecosystems

Unfortunately, too few universities are so far prepared to educate students to become specialists in horizontal research, i.e. train them to acquire an academic ability to piece together the fragments that vertical research brings forth from its deep trenches.

Thor Heyerdahl 1978

Fundamentals of Aquatic Ecosystems

EDITED BY

R.S.K. BARNES

Fellow of St Catharine's College
Lecturer in Aquatic Ecology
University of Cambridge

K.H. MANN

Professor, Department of Biology
Dalhousie University, Canada

BLACKWELL SCIENTIFIC PUBLICATIONS

OXFORD   LONDON   EDINBURGH

BOSTON   MELBOURNE

© 1980 by
Blackwell Scientific Publications
Editorial offices:
Osney Mead, Oxford OX2 0EL
8 John Street, London WC1N 2ES
9 Forrest Road, Edinburgh EH1 2QH
52 Beacon Street, Boston, Mass., USA
99 Barry Street, Carlton
    Victoria 3053, Australia

First published 1980
Reprinted 1982

Printed in Great Britain at
The Alden Press, Oxford

DISTRIBUTORS

USA
    Blackwell Mosby Book Distributors
    11830 Westline Industrial Drive
    St Louis, Missouri 63141

Canada
    Blackwell Mosby Book Distributors
    120 Melford Drive, Scarborough
    Ontario, M1B 2X4

Australia
    Blackwell Scientific Book
    Distributors
    214 Berkeley Street, Carlton
    Victoria 3053

British Library
Cataloguing in Publication Data

Fundamentals of aquatic ecosystems.
    1. Aquatic ecology
    I. Barnes, R S K
    II. Mann, Kenneth Henry
    574.5′263   QH541.5.W3
ISBN 0-632-00014-7

Contents

Contributors

R. S. K. BARNES   St Catharine's College and Department of Zoology, University of Cambridge, Cambridge, UK

G.E. FOGG   Department of Marine Biology, University College of North Wales, Menai Bridge, Anglesey, UK

F.R. HARDEN JONES   Ministry of Agriculture, Fisheries and Food Fisheries Laboratory, Lowestoft, Suffolk, UK

R.N. HUGHES   Department of Zoology, University College of North Wales, Bangor, Gwynedd, UK

K.H. MANN   Department of Biology, Dalhousie University, Halifax, Nova Scotia, Canada

W.E. ODUM   Department of Environmental Sciences, University of Virginia, Charlottesville, Virginia, USA

T.R. PARSONS   Institute of Oceanography, University of British Columbia, Vancouver, British Columbia, Canada

L.R. POMEROY   Institute of Ecology, University of Georgia, Athens, Georgia, USA

J.M. TEAL   Woods Hole Oceanographic Institution, Woods Hole, Massachusetts, USA

G.M. WOODWELL   The Ecosystems Center, Marine Biological Laboratory, Woods Hole, Massachusetts, USA

Preface

This book stems from the conviction that as scientific knowledge accumulates and becomes compartmentalized into many specialist disciplines, so it becomes increasingly difficult for the student and specialist alike to appreciate the form of the whole. This is as true in aquatic ecology as it is in other fields. Students may be taught separate courses in 'marine biology', 'freshwater biology', 'ecological theory' etc. yet, although convenient, this fragmentation obscures the essential unity of many of the processes taking place. Although somewhat unfashionable, synthesis is as valid an academic activity as analysis, nowhere more so than in the operation of the key processes of natural ecosystems.

Fundamentals of Aquatic Ecosystems is therefore an attempt to provide the undergraduate with a text concentrating on the common, fundamental features of all aquatic systems. Material relating to lakes, large ponds, lagoons, rivers, inland seas, estuaries, coastal seas and open oceans is integrated in an account based on the pattern of the food web, on productivity and on other aspects of ecosystem structure and function. It need hardly be emphasized that, on a world scale, some of these systems are more widespread, more important, and better known than are others, and hence equality of mention has not been attempted. Most chapters essentially treat one stage in the production process and describe, for that 'subsystem', the patterns of productivity, of nutrient flux, and of diversity and stability, with emphasis on the factors influencing these features and governing their magnitudes.

By restricting coverage to aquatic systems, we are not seeking to maintain a marked distinction between terrestrial and aquatic processes, indeed it is evident that all ecosystems conform to the same basic ecological 'laws'. Nevertheless, the processes and patterns characterizing aquatic systems do differ quantitatively, and in some cases qualitatively, from those typically to be found on land, and it can be argued that aquatic systems are more in need of a treatment such as that attempted in the following pages because of the greater tendency of research and its dissemination to fragment into specific habitat-based disciplines.

This book has been written for the undergraduate and therefore references to the literature have largely been confined to the 'further reading' or 'key article' type. We would like to think, however, that the novel approach which we have adopted will also make the book of some value to the postgraduate student and to our colleagues.

Finally, we wish to record our appreciation of the help which we have received from friends and colleagues, particularly from Robert Campbell of Blackwell Scientific Publications who was a pillar of logistic support.

R.S.K. Barnes   *K.H. Mann*
Cambridge   *Halifax, Nova Scotia*

Prologue

R. S. K. BARNES & K. H. MANN

This book is concerned with the nature of aquatic ecosystems, and it may be helpful to the reader with little ecological training to start with a brief description of some of the characteristics and properties of such systems, particularly of those which will be mentioned frequently in the following pages.

Green plants fix energy from the sun by the process of photosynthesis. The energy thus fixed is incorporated into various carbon-containing compounds and is, in part, subsequently dissipated by the respiration of the plants themselves and of those organisms which may consume them. Since this photosynthetically-fixed energy is ultimately the source of all the food available to consumers in the system, the food-generating plants are termed the *primary producers* and that which they produce is the *primary production*. The amount produced per unit time is then the *primary productivity*. Besides requiring light energy, photosynthetic plants need various *nutrients* (compounds containing nitrogen, phosphorus, etc.) which they take up from the environment and incorporate into their tissues, thereby partially depleting the environmental pool of these essential substances.

Consuming species may be classified on the basis of their diet as *herbivores*, *omnivores*, *carnivores*, etc. and, together, their growth and multiplication create the *secondary production* and *secondary productivity*. Consumers of the basic plant material are often termed *primary consumers*, and those capturing the herbivores are then the *secondary consumers*. Because of respiratory energy losses, secondary production will always be less than primary production, and, for example, that of carnivores will be less than that of herbivores.

Not all the energy content of green plants is directly available to herbivores, however; some plants may die uneaten and some of the tissues of others may not be readily digestible. The tissues of dead plants and the organic substances remaining in animal faeces, animal corpses, etc. are available to such organisms as bacteria and fungi which can decompose dead organic matter (i.e. *litter* or *detritus*). By decomposing these complex organic molecules, the *decomposers* will also release some of the incorporated nitrogen, phosphorus and other elements and put them back into circulation. In turn, the decomposers and their detrital food source may be consumed by detritus feeders, which are themselves taken by carnivores.

The interrelations of eater and eaten in an ecosystem (*trophic* relationships) form a *food-web*, which will therefore comprise a *grazing food-chain* of living plants → herbivores → carnivores, and a *detritus food-chain* of detritus → detritus

1

feeders → carnivores. The amount of food available at any one time to, say, the herbivorous consumers will depend on the total amount of living plant material present at that time (the *biomass* or *standing stock*), but over any period of time the amount of food available will depend on the rate at which it is being produced, i.e. on the productivity.

Different ecosystems or different parts of any one system will vary in the proportion of the total fixed energy which passes along the grazing as opposed to the detritus food-chain; in the magnitude of the biomass and productivity which they support; and in the relationship between the amount of production per year (annual productivity) and the amount present at any one time (biomass). Ecosystems also vary in the number of different types of organism which they can support, i.e. in their species diversity.

There is a tendency to assume that if an ecosystem has a large biomass of organisms present, it also has a high productivity. This is not necessarily true. There are several factors influencing the ratio of annual production to biomass (P/B). One is the longevity of the constituent organisms. Plants (such as trees) and animals (such as elephants) which accumulate a large biomass during their lifetime, are found to have a low P/B ratio when this is calculated on an annual basis. They can only flourish and build up their population where conditions remain favourable for long periods. In the process, the ecosystem of which they are part builds up a large biomass, and a large proportion of primary production is used to maintain the respiration of that biomass. This explains the low production to biomass ratio.

In environments where conditions are less favourable for large, long-lived species, ecosystems tend to be dominated by fast-growing, fast-reproducing, short-lived species. These have an opportunistic strategy of producing rapidly when conditions are favourable. Less of the primary production is used for maintaining high biomasses over long periods, so the P/B ratio tends to be high. Biomass will be high at one time, low at another, and is characterized chiefly by its variability.

The two types of species described above have contrasting life-history strategies. The long-lived, slow-growers tend to build up their biomass to the maximum that the environment will support (denoted by K for 'carrying capacity'), and to keep it there. The opportunistic species tend to have a strategy involving a high rate of reproduction and hence of population increase (denoted by r). The two types of strategy have been described by Berry (1977) as follows:

'Some species are highly mobile and colonizing, with high reproductive rates but low competitive abilities so that their populations tend to be ephemeral; whilst others fluctuate comparatively little in numbers, and put their energy into competition and maintenance. These alternatives [or, more accurately, the two extremes of a single continuum] have been called respectively r and K strategies . . . An r-strategist lives in a variable or an unpredictable environment and is subject to recurring heavy mortality; selection will favour rapid development and early reproduction, high reproductive rates, and small competitive abilities. In contrast, selection for a K-strategist leads to slower development, a greater body size, and the ability to withstand both inter- and intra-specific competition, since such species usually live in complex communities near their equilibrium numbers.'

2

Organisms with large biomass (such as trees) tend to be associated with complex communities for several reasons. One is that the biomass itself tends to generate spatial complexity. One has only to think of examples such as forests, or coral reefs, to realize that within them there are many kinds of habitats providing for many diverse kinds of organisms. Another reason for the complexity is that these communities persist for long periods, giving ample time for colonization by, and development of patterns of coexistence among, many species of plants and animals.

Diversity may often refer simply to the number of species per unit area, but ideally it should also take into account the relative importance of the different species, as the following example should make clear. Two carpets may each be constructed from threads of six different colours: one with one thread each of blue, green, brown, red and purple, and 10^6 yellow threads; the other with 1.6×10^5 threads of each colour. Both carpets have approximately the same total number of threads and both have threads of six different colours, but one will effectively be a uniform yellow colour whilst the other will have a diverse, multi-coloured appearance. A diverse system is therefore one in which not only are there more species, but in which the total number of individuals is more equitably distributed between the component species (the competitive situation is more balanced).

The last concept which we need to introduce in this Prologue is *stability*. An ecosystem is stable if it does not change much from day to day or from year to year, or if it changes only in a regular, predictable manner (like a sine wave). Stability, however, may arise or be consequent on either of two very different states. A 'house' built of playing cards is 'stable' provided that nothing disturbs it — *constant* would be a more apt term—and some ecosystems may be stable because of the absence of perturbing forces. Others are 'stable' in spite of perturbation; these are stable in the sense that a boxer's punch-bag is stable. If displaced they return to their previous state in a series of damped oscillations; they are *dynamically stable*. In ecosystems, these two states may well be mutually exclusive (in their extreme cases): dynamically stable systems show little constancy, whilst constant systems possess little dynamic stability (after all, their components have had no reason to evolve it!). In nature, constant systems are often relatively diverse, are dominated by K-strategists with low production: biomass ratios and are subject to a constant or predictable climate. Dynamically stable systems, on the other hand, are species-poor, are dominated by r-strategists with high production: biomass ratios and are subject to unfavourable, unpredictably fluctuating climates.

1

The Unity and Diversity of Aquatic Systems

R.S.K. BARNES

1.1   Introduction

Nature does not provide the ecologist, or anyone else, with discrete entities or patches: the natural world is a continuum. Man, however, has imposed artificial categories on nature, recognizing such things as species, oceans, ecosystems, populations and parasites, to name but a few. Indeed, were he not to do so, all enquiry and even communication would cease; yet these abstractions can have no objective existence and they are inherently undefinable. The differentiation of habitats into 'aquatic' and 'terrestrial' is one such characterization which on analysis will not—so to speak—hold water. Swamps and marshes, whether marine or freshwater, are clearly intermediate in character, and, for example, small temporary ponds or streams in a wood have very little claim to be considered entities at all, being merely wetter-than-average regions of that woodland. Somewhat larger bodies of water do, however, show a number of properties which differ in important details from those typically found on land.

In practice, man carries subdivision of the environment much farther than this. A glance along the ecology shelves of a bookshop will reveal that the aquatic habitat is normally apportioned between fresh, brackish, marine and inland salt ('athalassic saline') waters, and that freshwaters may be subdivided into running-water and standing-water systems. It is the burden of this book that, their apparent differences notwithstanding, the various types of aquatic system which have been recognized do all exhibit a fundamental unity. It is then the purpose of this introductory chapter to sketch the general nature of this underlying pattern, and to outline the more important parameters effecting the variations on that theme displayed by the individual aquatic habitats.

1.2   The unity of aquatic systems

Characteristically, and regardless of precise habitat-type, any aquatic system can be divided into three spatial compartments within which different processes are located: the pelagic community of the water mass; the benthic community living in and on the underlying sediments or rock; and, in shallow regions, the fringing communities dominated by emergent or submerged plants. For descriptive purposes, the pelagic compartment can be subdivided into the planktonic* community suspended in the water, and the nektonic* assemblage of larger,

*As Hutchinson (1967) points out, 'planktic' and 'nektic' are etymologically more correct, although 'planktonic' and 'nektonic' are hallowed by decades of use. 'Benthonic', however, is to be avoided.

more mobile organisms which can swim through it. Most aquatic systems support all four communities (plankton, nekton, benthos, and fringing), although their relative importance varies widely and some habitats may lack one, or more rarely two, compartments; for example, a true planktonic community may be absent from fast-flowing rivers.

In the light of the above introduction, it will be evident that these different communities will grade each into the others (and a host of terms such as 'benthonektonic' and 'nektoplanktonic' have been coined in a vain attempt to categorize intermediates), but for practical purposes the various communities or compartments still serve a useful analytical function when treated as semi-isolated subsystems. We will now look briefly at the nature and attributes of each community in turn.

1.2.1   PLANKTON

The plankton comprises all those aquatic organisms which drift passively or whose powers of locomotion are insufficient to enable them to move contrary to the motion of their inhabited water mass. Of course, this is not to suggest that plankton are necessarily incapable of moving within a given water mass; motile species may move vertically in a laterally-flowing current in much the same way that one may move up and down stairs in a double-decker bus without affecting one's movement relative to the ground. Such vertical migration provides, amongst other advantages, a means by which aquatic organisms can avoid unfavourable conditions and change water masses (since currents at different depths may flow at different speeds and in different directions) (see Chapters 3 and 9).

By definition, then, the plankton will include all those organisms suspended in the free water. Biological tissues are denser than water and hence only particles with large surface areas in relation to their volumes, and with slow sinking speeds, are likely to remain suspended. Small animals may counter this sedimentation by swimming upwards; large animals may also drift within currents, but since they are usually capable of active and powerful swimming, they will (again by definition) form part of the nekton. Therefore, most organisms belonging to the planktonic community will be small, although their size range does extend through several orders of magnitude and this has been utilized to form size-based classification systems. No system has been adopted universally (in part because the plankton attains a larger maximum size in the sea than in freshwater), but a representative one is: $<5~\mu$m, the largely bacterial 'ultra-plankton'; 5–50 μm, the largely algal 'nannoplankton'; 50–500 μm, the 'micro-plankton' of algae and animals; 500–2000 μm, the largely animal 'macroplankton', and $>2000~\mu$m, the animal 'megaplankton'. The systematic affinities of the different size classes are usually indicated by micro*zoo*plankton, micro*phyto*-plankton, etc.

As in the other communities (with the exception of the entirely animal nekton), primary producers, primary and secondary consumers, and decomposer organisms are all represented in the plankton, although it is the photosynthetic primary producers which are of especial significance in the ecology of many aquatic habitats: the planktonic subsystem is pre-eminently the site of production of those food materials directly available to grazing consumers. Almost all groups of algae have planktonic representatives and these, together with photosynthetic bacteria under certain conditions, fix carbon in those regions of aquatic systems illuminated by light in intensities in excess of about 0.002 Ly/min. Small size in an organism is also generally indicative of high productivity and low standing stock,* and this is true of the phytoplankton. In a year, phytoplankton produce an average of some 15–45 times their standing biomass, and therefore biomasses of usually well below 50 g dry wt./m² of surface may produce more than 1 kg dry wt./m²/yr. In many aquatic habitats this production is the major store of energy fueling the whole system, and hence a marked contrast with terrestrial situations results: on land, a reasonable average plant biomass: consumer and decomposer biomass ratio would be 1:0.001; in a phytoplankton-dominated aquatic system this ratio could be in the order of 1:20.

Planktonic algae are grazed by a variety of larval and adult zooplankton, amongst which the filter-feeding Crustacea are normally dominant, and these primary consumers are in turn eaten by predatory members of the plankton. The small size of the consumers again reflects high productivities in relation to small biomasses. The proportion of the phytoplankton production passing along this grazing food chain varies with a complex of factors, which include the nutrient status of the water. In areas rich in nutrient substances, algal production may be very large and much of it may go ungrazed. Dead and dying algal cells are decomposed *in situ* by the heterotrophic bacterioplankton, and although much of the detritus so created may sink to the benthic regions and thereby partly fuel that system, an increasing number of 'herbivorous' zooplankton are being shown capable of subsisting in whole or in part on suspended detritus. In addition, many of the planktonic grazers are inefficient feeders in that their faecal material is rich in undigested algal tissues. Being bound into a comparatively heavy faecal pellet, this fraction of phytoplankton production also sinks towards the benthos, being partially remineralized *en route* by bacteria. Thus the planktonic community exports materials to the benthos, although remineralization in the water column will recycle some nutrients; and it also exports material to the other pelagic community, that of the nekton.

Before considering the nekton, however, mention may be made of a specialized variant of the plankton which inhabits the air/water interface of aquatic habitats. The pleuston (the microscopic members of which are sometimes termed the neuston) span the same size range as the plankton but they have been much less intensively studied. Except in a few shallow, sheltered habitats, their contribution to the food-web is likely to be small.

---

*Blueweiss *et al.* (1978) have documented the relationships between body size and a number of processes of ecological relevance. The intrinsic rate of population growth (r_{max}) per day, for example, is inversely related to adult size according to the expression $r_{max} = 0.025W^{-0.26}$, where W is weight of organism in grams. Thus for an organism weighing 10^{-15} g, r_{max} per day is 200, whilst for a 100-kg organism it is 0.001. Or, expressed another way, an organism weighing 10^{-15} g could, assuming no constraints on growth, produce new tissue at a rate of 20 000% of its own weight per day, whilst one weighing 1 kg could only increase its weight by 0.4% per day. The 10^{-15} g organism would also pass through over 200 generations in 1 day; the 1-kg organism would take about a year to reach maturity.

7

1.2.2 NEKTON

The nekton comprises all those swimming consumers that constitute the middle and top trophic levels of aquatic systems. Fish are dominant, although other vertebrates (including some temporarily nektonic birds) are often important, and a few groups of invertebrates (especially cephalopod molluscs) may form a significant component in some habitats. Mobility is often a function of size and hence nektonic animals are relatively large. They are therefore usually long-lived and slow-growing, and accordingly their annual production is usually less than their standing biomass. Number of offspring produced per reproductive period is also often inversely correlated with size, particularly on land, but although there are several aquatic examples of a similar trend (e.g. whales), several fish have attained large size and long life, whilst still spawning prolifically. Amongst animal species, fish show the greatest change in size during their life; the eggs of cod (*Gadus morhua*), for example, are of the order of 1.5 mm diameter, whereas the adult may exceed 1.5 m in length. Some fish do produce a small number of eggs to which a degree of parental care is afforded, but the generally high fecundity of fish can be seen as a necessary response to the high mortalities to which small planktonic eggs and larvae will be exposed.

The mobility of the nekton enables them to remain in zones of high productivity in the face of currents which would otherwise carry them away, and it also permits them to migrate from one area to another and thereby select favourable habitats in which to spend different periods of their lives. A single nektonic species may be capable of feeding in all the various communities within a given aquatic habitat. Indeed, species like the eel (*Anguilla*) may be capable of exploiting all the different types of aquatic habitat at one time or another (its larval life is spent, successively, in the deep sea, in the surface waters of the open ocean, and in coastal and estuarine regions; whilst, when adult, it may inhabit ponds, lakes, rivers, lagoons, brackish ditches, saline lakes and inland seas): the eel is a living symbol of the unity of aquatic habitats! The fact that all food-webs of habitats large enough to support fish terminate in the nekton is intimately associated with their size and freedom of movement.

1.2.3 FRINGING COMMUNITIES

In marginal and other shallow regions of aquatic habitats, a second input of photosynthetically-fixed materials is provided by the fringing flora. This flora may take several forms:

1 Dense stands of vegetation of basically terrestrial ancestry rooted, and sometimes submerged, in the water (although frequently with a large part of their biomass extending into the air, either permanently or when uncovered by tidal water movements).

2 Somewhat equivalent stands of essentially aquatic macroalgae, temporarily exposed to the air or permanently submerged in shallow water (several species in this and the preceding category have leaves or fronds floating at the air/water interface).

3 Simple or colonial microalgae forming mats on (or otherwise attached to) the bottom sediments or rock; or inhabiting the interstitial spaces between sand

grains for example. The microalgae may be grazed by benthic animals or may be consumed by deposit feeders, but the larger macrophytes forming such stands as reedbed, mangrove-swamp or kelp-forest are generally not directly consumed by aquatic animals.

The importance of these larger plants to the aquatic systems which they fringe lies in the smaller epiphytic algae for which they provide an attachment site (their 'periphyton'), and in the litter and detritus which, on their decomposition, they supply to the water and to the sediments. Terrestrial species often consume those parts of semi-aquatic plants projecting out of the water and some macrophytes may be browsed when at the sporeling stage, but such utilization generally accounts for considerably less than 10% of the energy flow through most fringing plants. The precise proportion of the net primary production of macrophytes available only after its decomposition by fungi and/or bacteria is a controversial issue, as is the nature of the materials actually assimilated by detritus feeders (see below). The picture is complicated by the fact that several rooted angiosperms translocate potential food materials into their below-ground structures before shedding leaves, etc.; whilst some animals—generally known as shredders—may obtain substances directly from leaves before leaching and decomposition have started. Neither is it always easy to distinguish between consumption of the macrophyte and of the more delicate epiphytes. It seems likely, however, that most of the grazing or browsing invertebrates associated with the fringing aquatic vegetation are removing the periphyton (Fig. 1.1).

Fig. 1.1. Stereoscan electron micrograph of the feeding tracks of mayfly nymphs (*Chloeon dipterum*) grazing on the periphyton on *Chara* sp. ($\times 750$). [From Allanson B.R. 1973, *Freshwat. Biol.* **3**, 535–542]

Primary production of this fringing compartment of the aquatic system is high, and often very high; usually much higher per unit area than that of the planktonic algae. If the average phytoplanktonic production of a given habitat is accorded a value of unity, then the average macrophyte production would lie

9

within the range 0.5–50 dependent on the climate and the precise nature of the environment. Of course, the biomass of macrophytes is also very high, and in fact production: biomass ratios seldom exceed 1:1. In contrast, the comparatively small standing stocks of the epiphytes produce up to some 30 times their biomass each year, so that the productivity of these visually insignificant components of the vegetation may, under favourable conditions, actually exceed that of their larger associates.

The production of the epiphytes and of the benthic microalgae is consumed *in situ*, as also will be some of the detrital production of the macrophytes, but plant debris is carried suspended in the water or as flotsam and some will be transported to other parts of the habitat or even to other habitats. This export may account for 50% of the net macrophyte production, and there is a detectable input of detritus derived from fringing stands of vegetation to the deep offshore waters of large aquatic systems like the ocean. The fringing plants can therefore be likened to, on the one hand, large detritus factories producing a water-borne export to other compartments and systems; and, on the other, to inert frameworks providing a large surface area for the attachment of small digestible plants (and animals) of high productivity.

1.2.4 BENTHOS

In shallow waters, benthic and nektonic grazers can therefore subsist on a diet of algae; in deeper waters, however, living plants will be absent and the benthos of such areas can only be supported by the rain of detrital material from the other aquatic communities. The benthos of regions below the depth of the illuminated zone is thus pre-eminently the site of the detritus food chain, of heterotrophic bacterial activity, and of nutrient regeneration.

Detritus is a composite term embracing decaying plant and animal matter, equivalent material in animal faeces, the decomposer microorganisms, micro- and meiofaunal protists, flatworms, copepods, etc.; and, in shallow regions, attached or otherwise associated algae. It therefore constitutes a whole microcosm and food-web; moreover, even one component, the ciliate protists for example, may in itself form a complex subsystem (Fig. 1.2). Experimental work, mainly using detrital analogues, has endeavoured to elucidate the nature of the components assimilated by detritus feeders, but to date the results have been confusing. Studies could be cited to emphasize the importance of any one component and the insignificance of all others, whilst, under laboratory conditions, some species can utilize them all if provided in pure culture, although with varying degrees of efficiency. By the time sinking detrital material has reached the bed in deep water, it is likely that most of the more easily digestible substances have already been decomposed by bacteria, and hence the bacterial biomass itself may be of overriding significance; but even if the bacteria and fungi are not directly assimilated, it is probable that their activity is required to convert the refractory substances remaining into utilizable food. The importance of the abundant and productive meiofaunal organisms is largely unknown. They have been considered to be the component through which most energy flow takes place, or perhaps 'into which' might be more appropriate since it has also been suggested that they represent a trophic cul-de-sac. Recent studies, however,

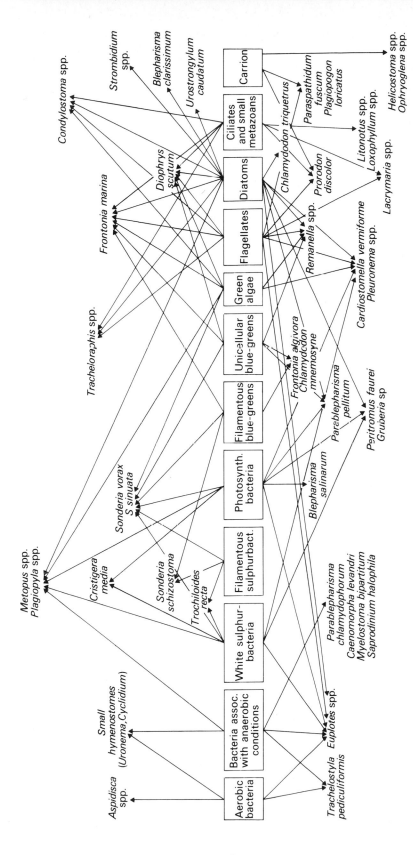

Fig. 1.2. Food web of the most common ciliate protists in estuarine sediments. [From Fenchel T. 1969, *Ophelia* **6**, 1–182]

have implicated them in the diet of several animals hitherto regarded as carnivores or consumers of debris.

Aquatic sediments are often rich in organic matter, so much so that most fossil fuels are nothing but the geologically processed accumulations from such systems. Yet many detritus feeders are fairly slow-growing and long-lived species, with large biomasses but low productivities, suggesting that food materials are not present to excess. What then can limit benthic production in otherwise favourable environments? It will be immediately apparent that the surfeit of organic matter cannot all constitute suitable food, and several workers have suggested that microbial productivity and colonization of detrital particles are limiting. Because decomposed organic matter contains only minute amounts of nitrogen and phosphorus—nutrients required in comparatively large amounts by bacteria—bacterial production and hence decomposition may be limited by nutrient shortage within the sediments, and this in turn will affect larger organisms higher up the detritus food chain.

Detritus for consumption is collected by a wide variety of techniques involving differing degrees of selectivity, but in general detritus feeders can be categorized by whether the material is in the water column or in and on the sediments. Regardless of strategy, many species are sessile or sedentary: a life style barely possible on land. Suspension feeders, which filter suspended particles from the water or trap sedimenting material before it has gained the bottom, are most characteristic of shallow regions (where planktonic food is also available) and areas of comparatively rapid water movement. Deposit feeders, which consume a specific fraction or the whole of the surrounding sediment, occur in greatest abundance in deeper areas and in soft sediments. Many species in both categories burrow or are buried in the bottom and thus comprise part of the 'infauna', although it is noticeable that in the absence of predators, as in parts of the deep sea for example, several otherwise infaunal groups are represented by species dwelling, exposed, on the surface of the sediment ('epifaunal').

These detritus-based communities support a number of resident invertebrate predators, and all trophic levels are consumed by benthic feeding nekton. In the deep stable regions of large aquatic systems, the benthic fauna is diverse, stable, although unproductive (when the physical environment permits the establishment of a benthos; see below). The levels of productivity achieved more or less parallel the pattern of productivity in the overlying water, with one important exception. In regions of very high productivity in the water column, the fall-out of organic matter on to the bottom may be sufficiently great for its microbial decomposition to deplete the bottom water of oxygen: the underlying areas may then be devoid of animal life.

Many groups of benthic animals have planktonic larvae. We have seen that food may be in relatively short supply in the benthos, but relatively abundant in shallow and surface waters where the photosynthesis powering the whole system is located. Many organisms, representing most systematic groups, indulge in habitat changes correlated with the requirements of different stages in their life cycle, such that the rapidly growing young develop in an environment more able to support this growth—the aquatic larvae of a number of terrestrial insects, the coastal and estuarine nurseries of young marine fish, and the breeding migrations of several birds are expressions of this phenomenon. The plankton-feeding larvae of benthic animals can, at least partially, be interpreted in this light. It has even been suggested that the permanently zooplanktonic organisms have been derived, by paedomorphosis, from such larval stages (see for example, de Beer 1958, Gould 1977); and that the reduction in the adult life-span (or lengthening of the larval life) of mayflies, stoneflies and other freshwater insects, and the neotenous larvae of some amphibians are other adaptations to spending the whole or as much of the life span as possible in areas of relative food abundance (areas with comparatively little community biomass supported by unit primary production).

The foregoing is a somewhat schematic outline of some of the more important features of the four compartments into which all aquatic systems can functionally and spatially be divided (and see Fig. 1.3). It is clearly based on a rather hypothetical aquatic system: one from which have been shorn all the modifications occasioned by the different types of physical environment to which aquatic habitats are subject. We must now consider these various modifications in their own right.

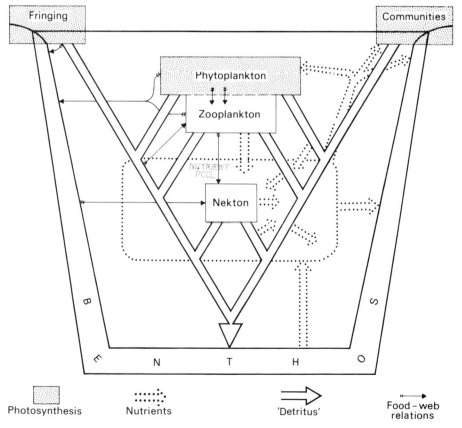

Fig. 1.3. Diagram showing the major interrelations of the four semi-isolated subsystems into which aquatic ecosystems can be divided.

13

1.3 The diversity of aquatic systems

Many of the factors creating the wide variety of aquatic habitat-types which may be distinguished act by altering the relative importance of the four compartments outlined above; others relate to the size and shape of the container in which the water flows or circulates; whilst perhaps the most important single factor concerns the extent to which the water is stratified.

1.3.1 STRATIFICATION

Water bodies stratify when stable density differences are generated, often as a result of surface heating (with the establishment of a thermocline), but sometimes owing to differences in salinity of the participating water masses (with the formation of a halocline). Such stratification provides a barrier to nutrient circulation. If wind is the major agency promoting mixing within the water body, wind-induced turbulence may have insufficient power to penetrate a deep and/or marked thermocline, thus leaving a stagnant and deoxygenated lower water mass. In tropical and subtropical areas, the constantly high energy input from the sun creates a warm surface layer permanently floating upon the colder water at greater depth (in both the sea and inland water bodies whether fresh or salt). In temperate regions, this stratification is a seasonal phenomenon, occurring only in the warmer months of the year.

A thermocline greatly reduces mixing between the productive surface layers (the region of nutrient fixation) and the zones beneath in which most decomposition is occurring (the region of remineralization). Faeces, corpses, etc. sinking out of the sun-lit zone will therefore drain the surface waters of nutrients, except insofar as they are released before passing through the thermocline; and if the thermocline is permanent, equally permanent nutrient shortage in the surface waters will result. The surface waters of large tropical systems will come into this category (Fig. 1.4) and can therefore support comparatively little production, except where wind-driven upwelling takes place along a coastline. Phytoplankton communities in such regions are adapted to remove nutrients from low external concentrations, however, and selection favours efficiency under the conditions of intense competition. Thus although primary production is low, relatively high community biomasses are often maintained per unit of this production, and the comparatively constant environmental conditions pertaining permit communities of great diversity to evolve. In contrast, in regions sufficiently cool for a thermocline never to become established, there are no barriers to mixing and so nutrients may be plentiful, but the energy input for photosynthesis will for much, if not all, of the time be insufficient to support much production. Temperate regions, of course, will oscillate seasonally between the two states, with winter injection of nutrients into surface waters and summer depletion of the entrapped stocks. Phytoplankton 'blooms' may then rapidly build up and die down, and a low diversity system with higher production to biomass ratios will be characteristic, as the environment is here less constant or predictable and production is less efficiently converted into community biomass in a system fluctuating between luxuriance and scarcity.

Wind is not such an important factor in the mixing of open oceanic water:

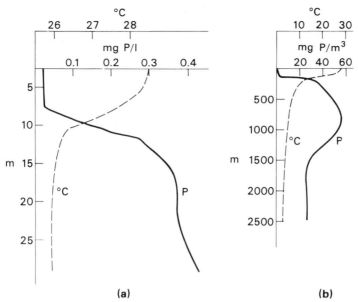

Fig. 1.4. Vertical distribution of temperature and phosphate in: (a) a tropical lake (Ranu Lamongan, Indonesia), and (b) a tropical ocean (the North Atlantic, 2–10°N & 40°W), showing the nutrient impoverishment of surface waters associated with a persistent thermocline. ['a' after Ruttner F. 1931, *Archiv. für Hydrobiol.*, Suppl. *8*, 197–454; 'b' after Seiwell H.R. 1935, *J. Cons. Int. Explor. Mer.* **10**, 20–32]

the circulation systems of these large water masses are generated by low-temperature-induced differences in density (a process in which the polar ice-caps are particularly important). Large-scale water movements keep oceanic water oxygenated right down to the sea bed, and the presence or absence of a thermo-cline has no bearing on the aeration process. Land-locked fresh or salt lakes and inland seas, however, rely heavily on wind-induced mixing to convey oxygen-rich water to the depths, and stratification will interfere with this move-ment. An equivalent phenomenon occurs in those arms of the sea separated from the parent water mass by a sill or shallow area at their mouth, e.g. many fjords and the Black Sea. The rate of exchange of bottom water within such inlets is limited, and to some extent the deeper waters of these systems form isolated ponds or lakes.

Except in the oceans, therefore, once a thermocline is established, oxygen in the bottom waters cannot be replenished. Microbial decomposition of the rain of organic matter uses the oxygen and, if the thermocline persists for long enough, exhausts it. On deoxygenation, first nitrates and then sulphates can be used as oxygen-donors, with, when sulphates are used, the liberation of poisonous hydrogen sulphide. Anoxia will therefore be a permanent feature of the bottom waters of deep lakes and inland seas in warmer parts of the world and a seasonal phenomenon in temperate zones.

In such situations, life below the thermocline is normally restricted to anaerobes equivalent to those dwelling in the reduced layers of all benthic sediments, including various chemosynthetic bacteria which can use methane, hydrogen sulphide, sulphur, carbon monoxide, hydrogen and other reduced compounds as inorganic substrates in the fixation of carbon. The non-microbial benthos will be restricted to those shallower benthic regions bathed by the oxygenated water mass above the thermocline. Thus apparently single aquatic systems in reality may, temporarily or permanently, comprise two separate systems one above the other: (a) an upper one in which all photosynthetic production and primary formation of detritus are located and to which the plankton, nekton and non-microbial benthos are largely confined; and (b) a lower, largely azoic sink in which nutrients accumulate in a reduced state, only slowly to diffuse back towards the surface, and in which anaerobic processes are responsible for processing the detrital fall-out. Deoxygenation will clearly be most marked when production in the surface water is large; nutrient-poor systems may support such low levels of production that its decomposition removes insignificant amounts of oxygen from the water and the bottom waters remain oxygenated.

Similar circumstances are found in some estuaries and several lagoons where salinity differences lead to the establishment of a halocline: this functions in a comparable manner to the example of the thermocline described above.

1.3.2   DISSOLVED SALTS AND NUTRIENT STATUS

Most practical classifications of aquatic habitats are based on the quantities of dissolved inorganic salts or other solutes. Yet whilst the salinity of the water greatly affects the nature of the fauna and flora of a given area, it plays little direct role in determining the form of ecological play enacted. Thus although more than twice as many classes of animals are represented in the sea than in freshwater, the diversity in terms of individual species does not differ appreciably. There are, of course, important differences which have a bearing on aquatic ecology: ice is a more widespread phenomenon in freshwaters; freshwaters are more variable in their pH, Eh, dissolved salts, oxygen concentrations, etc.; suspension in the water column is more difficult for the freshwater plankton; and so on. But in comparison to the other features under consideration in this section, differences between freshwaters and the sea dependent on their salinity are all minor.

Salinity is of more importance as a variable within the non-marine aquatic habitats, because it is there generally correlated with levels of primary production (Table 1.1). In hot, dry regions, lakes may be found in which the volume of freshwater input is more or less balanced by evaporation from the lake surface. Salts are therefore concentrated within the water mass, and a salt-lake or inland sea is formed, the salinity of which slowly increases with time. Above about 10–15‰ S., the fauna of such systems becomes impoverished, and the Dead Sea (226‰ S.) is so-called because of the absence of animal life. Even when the salinity of salt-lakes is equivalent to that of the sea, such animals as are present are of freshwater, not marine, origin; unless, like the Caspian Sea for example, the water mass has passed through a marine phase in its history. The ionic ratios of athalassic saline waters differ markedly from sea water (Fig. 1.5) and vary from one salt-lake to another; thus there are carbonate, sulphate and chloride lakes. Nutrients such as nitrates and phosphates are also concentrated by evaporation, and algal productivity is usually higher than in the most

16

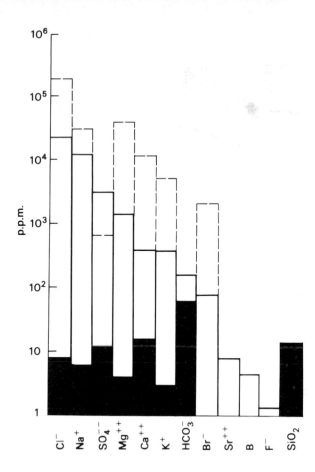

Fig. 1.5. Concentrations of major ions in sea water at 35‰ salinity (top of open bars), in world average river water (top of solid bars), and in the Dead Sea (top of dashed bars). [Modified, with additions, from Phillips A.J. 1972, In: R.S.K. Barnes & J. Green (eds.), *The Estuarine Environment*, Applied Science, London]

productive of freshwater lakes (Table 1.1): even the Dead Sea has a flourishing algal population and is highly productive.

This relationship between nutrient status and algal production is also evidenced by a number of other circumstances. Freshwaters, for example, usually contain higher nutrient concentrations than sea water, including of those complex humic acids leached from soil which appear to stimulate the production of various algae (Fig. 1.6). Therefore, estuaries, lagoons and coastal marine areas are more productive than the open ocean (Table 1.2), at least in part as a result of nutrient injection from rivers (see also pp. 19–20). Freshwaters, however, vary

Table 1.1. Typical ranges of phytoplankton biomass and productivity in relation to nutrient status and salinity in inland aquatic habitats. (After Likens 1975)

	Oligotrophic	Mesotrophic	Eutrophic	Saline
Net primary productivity (g dry wt./m²/yr)	15–50	50–150	150–500	500–2500
Phytoplankton biomass (mg dry wt./m³)	20–200	200–600	600–10 000	1000–20 000
Total phosphorus (ppb)	< 1–5	5–10	10–30	30–100
Inorganic nitrogen (ppb)	< 1–200	200–400	300–650	400–5000
Total inorganic solutes (ppm)	2–20	10–200	100–500	1000–150 000

17

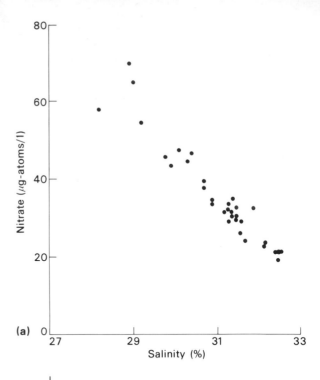

Fig. 1.6. Relationship between salinity and (a) nitrate (in Southampton Water on 23rd March 1970), and (b) humic acids ('Gelbstoff') (in the Baltic Sea). [From Phillips 1972, *loc. cit.*]

Table 1.2. The magnitude of aquatic productivity in different environments. [After, and from data in Whittaker 1975]

Habitat	Net primary productivity g dry wt/m²/yr		Secondary productivity g dry wt/m²/yr	Plant:animal productivity ratio
	range	mean	mean	1:
Open ocean	2–400	125	8	0.06
Upwelling areas	400–1000	500	27	0.05
Continental shelf	200–600	360	16	0.04
Algal beds/reefs	500–4000	2500	60	0.02
Estuaries*	200–3500	1500	34	0.02
Lakes and rivers	100–1500	250	5	0.02
Swamps and marsh	800–3500	2000	16	0.008
[Non-aqueous habitats	0–3500	760	0–20 (6)	0.003–0.02 (0.008)]

*excluding their fringing marshes.

18

markedly in their nutrient concentrations as these depend on the nature of the land over and through which they have passed. This is reflected by the correlation between the nutrient status of freshwater lakes and the levels of primary production which they support (Table 1.1); the nutrient-poor oligotrophic lakes show comparable features to the open waters of tropical systems (e.g. high species diversity, low production, and low primary production to community biomass ratios), and the nutrient-rich eutrophic lakes resemble the state described earlier for algal blooms.

Lakes may change their nutrient status relatively rapidly as the rocks and soils in their watersheds respectively erode and become leached, and as particulate materials accumulate in their basins. The direction of change—whether oligotrophic lakes evolve towards the eutrophic state, or vice versa—has, however, given rise to much argument, and the matter is further complicated by the fact that many of the 'textbook' lakes cited in the debate have been glacially modified. It seems almost certain that glacially created lakes are initially oligotrophic, but as soil develops in their watersheds more nutrients are washed into the system and they become more eutrophic. But the presence of a soil and vegetation cover protects the underlying rock from further erosion, and the soil then contains a finite pool of (some) nutrients which can only be depleted during further passage of time: a slow drift back towards oligotrophy would thereafter be the most likely course of events. Natural eutrophication can therefore be considered a glacial perturbation in an otherwise eutrophic → oligotrophic sequence. Both natural changes, however, are dwarfed by the input by man of nutrients into many natural systems, via sewage, detergents, etc., leading to cultural eutrophication. Comparatively few lakes remain oligotrophic near to centres of high human population density.

1.3.3 DEPTH AND SHORE-LINE DEVELOPMENT

A further factor related to stratification and nutrient status is depth, and particularly the proportion of the bed of a given system overlain by shallow depths of water. It is difficult to define 'shallow' with any degree of precision: for some purposes, 100 m is shallow; for others, 30 m is deep. The actual variables involved are the extent to which the benthos is illuminated and the degree to which the benthos is separated from the surface waters. Regions of the bottom receiving light intensities in excess of the 0.002 Ly/min mentioned earlier will support benthic photosynthesis, and in depths of less than some 20 m this may exceed phytoplanktonic production by up to a factor of ten; whilst in these shallow waters the benthos can feed directly on the plankton, and wind and current-induced mixing can reinject the products of decomposition back into the surface water layers. The efficiency of energy transfer and nutrient cycling will thereby be enhanced.

In addition, a fringing semi-aquatic plant community can develop in the still shallower littoral zones, and the contribution of this compartment to the total fixation of energy will, *inter alia*, therefore depend on the ratio of the shallow, immediately coastal areas to the total area of the system. Small water bodies and those with highly convoluted coastlines are likely to be (a) more productive, and (b) more highly dependent on detritus than larger systems and those with shore-lines plunging straight down to considerable depths, e.g. crater lakes and fjords. These features are also, in part, responsible for the pattern seen in Table 1.2; even in a large system like the ocean, benthic and fringing plants, which can only colonize some 0.1% of the surface area, probably contribute 10% of the total primary production and, since some 17% of the planktonic primary production of the seas is also coastal (discounting that from zones of upwelling), a total of more than one quarter of the marine photosynthetic production is contributed from the shallows.

1.3.4   SHAPE OF THE CONTAINER—CHANNEL OR BASIN

It will not have escaped notice that so far I have concentrated on aquatic systems occupying basins and have neglected rivers—unidirectional, running-water systems flowing along narrow, parallel-sided troughs. In terms of the volume of water which they contain, rivers are the least important of aquatic habitats (Table 1.3), but they pass over 30 times their contained volume to the sea each year, and their abundance and role in terrestrial ecosystems and agriculture make them of great importance to man.

Table 1.3. Percentage of the earth's water occurring in different aquatic systems. [After Wetzel, 1975]

Oceans	97.6
Ice	2.1
Exchangeable ground- and soil water	0.3
Freshwater lakes	0.01
Saline lakes	0.01
Atmospheric water vapour	0.001
Rivers	0.0001

In fact, rivers are not as different from other aquatic systems as might appear at first sight. Their upper reaches show several parallels with rocky marine shores, for example, and their silty lower reaches approximate to lagoons, estuaries and marine or freshwater 'backwaters'. The characteristic semi-aquatic fringing vegetation, nekton, benthic algae, predators, and browsing, suspension and deposit feeders are all present, although the unidirectional water flow does result in a comparatively insignificant planktonic component, at least in the upper reaches of some rivers. Rivers, however, almost always do support a 'potamoplankton'. In the fast-flowing upper reaches this may be dominated by typically benthic diatoms swept into suspension, but where water flow is less rapid a true plankton occurs in which diatoms and rotifers are especially important; densities may approach 100 million and 1 million per m^3, respectively. In contrast to standing-water systems though, the crustacean plankton is often relatively unimportant, and the phytoplankton biomass usually greatly exceeds that of the zooplankton: these are probably both reflections of the selective pressure exerted by the rapid flushing times of running-water systems in favour of the faster turnover rates of small organisms.

The hydrographic characteristics of rivers and their estuarine mouths also ensure that they will be essentially through-put systems, receiving and being dependent on a large import of organic matter from outside their confines (leaves from overhanging trees, materials leached from the surrounding land, etc.) and transporting much of their own production to adjacent systems, especially during the peak flow periods of spates. This makes it particularly difficult to consider a river as even a semi-isolated unit.

The opposite of the turbulent conditions near the head of a river is provided by habitats in which little vigorous water movement occurs, often as a result of the protection from wind afforded by the encircling land. I have argued elsewhere (Barnes 1974) that estuaries support an essentially marine fauna which tolerates the lowered and fluctuating salinity regime to the best of its varying ability in order to inhabit the food-rich muds. Soft sediments and the accumulation of finely particulate detrital materials are both consequent on shelter from wave and/or current action. Deposit feeding is a characteristic of sheltered habitats, whilst suspension feeding is typical of more exposed (and therefore more rocky) regions. Lagoons, like estuaries, are detritus-based systems dominated by marine organisms (Barnes 1980); but some of these epitomes of shelter may support large numbers of swimming insects, in marked contrast to most estuaries. This may be another consequence of placid water movement.

1.3.5 ISOLATION AND EPHEMERALITY

The world's continental blocks (the 28.5×10^6 km^2 of the continental shelves and their 149×10^6 km^2 emergent portions) occupy a finite 35% of the global surface. The blocks may move, subdivide and reassemble, but their total area remains constant. The nature of their surface, however, and the habitats which they provide can change drastically over relatively short periods of time, whilst the remaining 65% comprising the ocean basins is effectively a permanent habitat: the shape of the various oceans has changed and will continue to change, but this notwithstanding the oceanic flora and fauna have experienced continuity of their geographical habitat for many millions of years. Such is not true of the other aquatic habitats. Lakes and ponds, in particular, are islands scattered through an environment hostile to most aquatic organisms, and islands moreover with geologically a very short life.

Effective powers of dispersal are therefore a prerequisite for long-term freshwater life, and it may be no coincidence that many freshwater organisms can either survive to a limited extent or spend part of their life cycle on land, or else form resting bodies which enable them to survive periods without water *in situ* and to be passively dispersed through the air. The feet of amphibious birds are often invoked (apocryphally?) to explain other distribution patterns. The ability to form highly resistant resting bodies is most marked in the small organisms inhabiting the tiny temporary 'ponds' which occur in the axils of leaves, tree hollows, farmyard ruts, etc. These microcosms may form complete aquatic systems lacking only a nekton; they may even be used by some amphibians and crabs as habitats for their larvae.

Of necessity, however, their powers of dispersal will be limited in comparison to the distances separating some freshwater habitat types, particularly when the intervening environment is the sea. The freshwater invertebrates of oceanic islands therefore are often derived from the surrounding ocean, and the fauna colonizing islands from which glaciers have retreated has been suggested to be

21

of marine ancestry. Such species may be displaced by more truly freshwater competitors (usually of immediate terrestrial ancestry) if and when they finally arrive. This pattern of local extinction and colonization ensures that many aquatic communities will be evolutionarily young. This state may be contrasted with that shown by the deep-sea fauna, with its very high levels of diversity, its great stability and the poorly developed dispersal powers of its species.

1.4   Conclusion

Aquatic systems undoubtedly comprise a diverse assemblage of habitats, ranging from reedy lakes to the open ocean, and from mountain streams to coral reefs. This habitat diversity is reflected in their faunas and floras: several habitats could be cited without a single species or even a taxonomic order or class in common. I have tried to indicate that this apparent heterogeneity is illusory, however, if one considers the fundamental nature of the ecological systems formed by these varied species. All aquatic systems can be divided into the three semi-isolated compartments of the pelagic (planktonic and nektonic), fringing and benthic communities, each performing its characteristic functions regardless of precise habitat type. The plankton supplies and partially uses the living plant food; the fringing vegetation provides living periphyton, microalgae and associated small animals to the benthos, and detritus to the whole system; the benthos processes and remineralizes the detritus and debris; whilst the nekton consumes the organisms of the plankton and benthos, and is itself exploited by man. The differences manifested by the various habitats are not qualitative but are of degree, dependent on such variables as the shape of the container, the extent to which the contained water is mixed, and so on. Even rivers may have their diatom and rotifer plankton, and permanently stratified lakes their decomposers in the benthos.

This has been a very general introduction to the basic nature of aquatic systems; more detailed information on the individual habitats included under this head may be found in such works as Zenkevitch (1963), Hutchinson (1967, 1975), Hynes (1970), Remane and Schlieper (1971), Bayly and Williams (1973), Menzies et al. (1973), Brinkhurst (1974), Perkins (1974), Wetzel (1975), Whitton (1975), Cushing and Walsh (1976), Barnes (1977), Parsons et al. (1977) and Barnes (1980). Nevertheless, it should now be possible to go on to examine in greater depth the properties of the various component parts of the system that occupies 72% of the earth's surface and a much larger percentage of the volume of the biosphere (Table 1.4). To comprehend fully the nature of the aquatic

Table 1.4. Comparison between terrestrial and aquatic production (partly from data in Whittaker 1975): World totals.

	Surface area (10^6 km^2)	Approx. volume of biosphere* (10^6 km^3)	Net plant production (10^9 t/year)	Animal production (10^6 t/year)
Terrestrial	145	14.5	110.5	867
Aquatic	365	1445	59.5	3067
Ratio	1:2.5	1:99	1:0.54	3.54:1

*based on average sea depth of 4000 m and assuming an average terrestrial inhabited zone 100 m deep.

22

ecosystem, however, it is necessary first to forget some of the preconceptions natural to a terrestrial organism. Although, as stated earlier, all ecological systems—however aqueous or arid—conform to and reflect the same general principles and processes, aquatic systems are not just wet versions of the more familiar land. The patterns of expression of these principles and processes may be very different. The land, for example, is very obviously plant-dominated; with a total plant biomass dwarfing that of the other components of the ecosystem, and with individual plants often achieving large size and forming a major structural part of the habitat. In contrast, through much of most aquatic environments—whether coral reef or open ocean—animals are most in evidence; individual plants being minute, inconspicuous and of small total standing stock. This difference is a result of the disparate strategies appropriate to aquatic and terrestrial plants in respect of light, nutrients and gravity, with consequent effects on optimal sizes. The size difference has other ecological consequences. Land plants possess much structural carbohydrate material which is resistant to animal enzymes; aquatic plants, however, are both individually ingestible in their entirety and more digestible. Thus the dynamics of herbivore/plant interactions are necessarily different in the two types of environment, and the efficiency with which plant biomass and production are converted into animal biomass and production will also differ. Although they only account for some 35% of the world's plant production, aquatic systems contribute nearly 80% of its animal production (Table 1.4).

In their gross structure, both aquatic and terrestrial environments comprise a firm rock or sediment substratum overlain by a mobile and circulating fluid, but the relative biological exploitation of these compartments is dissimilar. Perhaps the simplest way to gain an appreciation of this difference is to look out of your window on a dark night and imagine the ground to be devoid of plant life, but the air to be inhabited (permanently) by organisms ranging in size from $<1\ \mu m$ to >30 m. Some of the animals are being blown willy-nilly by the wind, others—some larger than the largest dinosaur—are flying strongly whilst sieving out or capturing individually the smaller species. On the ground are diverse other communities of organisms existing by consuming the soil microbes and the rain of decomposing fall-out. It is cold and dark, although high in the sky, beyond the range of sight, would be a thin layer warmed and illuminated by the sun. It alone might be coloured a faint green. Such would approximate the scenario of 51% of the world's animal biomass.

POSTSCRIPT

Whilst this book was in production, two papers particularly relevant to themes under discussion appeared. In the first, Høpner Petersen and Curtis (*Dana* **1**, 53–64; 1980) point out that a relatively greater proportion of phytoplankton production is channelled through the benthos of marine shelf systems in high latitudes than in the tropics, and they discuss some of the consequences of this difference. In the second, Cammen (*Oecologia Berl.* **44**, 303–310; 1980) derives a relationship between a further biological variable and size (see p. 7): he shows that regardless of the nutritive status of the food, the rate of ingestion of organic matter, C (in mg dry wt per day), is related to body weight, W (in mg dry wt), by $C = 0.381\ W^{0.742}$.

2

Phytoplanktonic Primary Production

G.E. FOGG

2.1   Introduction

The chemical elements involved in living processes, being virtually indestructible, are recycled within the biosphere. The energy necessary to drive these processes, on the other hand, is derived from external sources, which may be either radiant or chemical, and it flows through the biosphere to be dissipated finally and irrevocably as heat. The cycling of elements and the flow of energy being linked at many points, their rates are interdependent and, although the relationships are complex, show a general positive correlation with each other. In this chapter we are concerned with primary production, that is, with the input of externally derived energy, into the planktonic ecosystem, but in discussing this it is inevitable that aspects of mineral nutrition should be referred to if a proper understanding of the functioning of the ecosystem is to be obtained.

2.2   Processes contributing to primary production

The most important input of energy into the planktonic ecosystem is by photosynthesis. This is the synthesis of organic matter from water, carbon dioxide and mineral salts as carried out in sunlight by chlorophyll a-containing plants. It may be described by the equation:

$$CO_2 + H_2O \xrightarrow{h\nu} (CH_2O) + O_2 \qquad \qquad (Equation\ 2.i)$$

in which (CH_2O) is an approximate empirical formula for organic products, the potential chemical energy of which has been derived from radiant energy. Photosynthesis need not, however, necessarily start from the inorganic form, carbon dioxide. If a suitable organic carbon source, such as acetate or glucose, is available cell material may be synthesized directly from this using energy supplied by the photosynthetic mechanism. Thus a substance already having high potential chemical energy is converted into other substances of high potential chemical energy at the expense of radiant energy. This process, photo-assimilation, is well attested by laboratory experiments; it presumably occurs, for example, in water polluted with sewage, but it is not clear to what extent it goes on under natural conditions. There is yet another variant of photosynthesis carried out by bacteria, which contain chlorophylls other than chlorophyll a, and by chlorophyll a-containing algae under special conditions, which depends on the availability of particular inorganic or organic hydrogen donors and in which oxygen is not evolved. For example, green sulphur bacteria carry out photosynthesis according to this equation:

$$CO_2 + 2H_2S \xrightarrow{hv} (CH_2O) + H_2O + 2S \qquad (Equation\ 2.ii)$$

Bacterial-type photosynthesis is only possible under anoxic or microaerophilic conditions. Thus it does not occur to any appreciable extent in the open waters of seas or lakes or in rivers but it does become important in waters (either salt or fresh) in which extensive decomposition of organic matter is occurring. Chemosynthesis, in which organic compounds are synthesized using energy released by the oxidation of inorganic materials, as for example happens when nitrifying bacteria grow at the expense of the oxidation of nitrite to nitrate, may also contribute to primary production. In addition, it must be remembered that photosynthesis and growth by plants produce conversions of inorganic substances which involve energy changes. Thus the assimilation of nitrate-nitrogen involves its reduction to ammonia; and storage of orthophosphate as polyphosphate, which is common among algae, also represents an accumulation of potential chemical energy. Strictly speaking, since primary production is defined as input of potential chemical energy into the ecosystem, all these processes—photoassimilation, bacterial photosynthesis, chemosynthesis and conversions of inorganic substances—ought to be taken into account. In practice in most oxygenated environments the type of photosynthesis defined in Equation 2.i) is predominant and since it is not usually determined to an accuracy much better than $\pm 10\%$, contributions to primary production from these other processes may be disregarded. In assessing primary production in anoxic environments bacterial photosynthesis must obviously be taken into account. The assessment of primary productivity in waters receiving organic pollution is peculiarly difficult; photosynthesis, photoassimilation and heterotrophic assimilation may simultaneously all be contributing substantially to the synthesis of new cell material so that there are problems in determining the utilization of radiant energy and there is the difficult question of whether input of potential chemical energy derived from organic matter from outside the ecosystem may properly be described as primary production, as perhaps it should be on the definition adopted here.

2.3   The measurement of primary production

2.3.1   GENERAL CONSIDERATIONS

Thermodynamically the most desirable measure of primary production would be change in free energy but, at present at least, this cannot be done easily enough to be of practical use. If, as is usually so, we require large numbers of determinations for comparative purposes we are limited to assuming that oxygenic photosynthesis is the predominating contributor to primary production and to measuring changes in the amount of one of the substances involved, e.g. oxygen or carbon. With terrestrial plants a simple method of doing this is to determine the dry weight of material accumulated during the growing season. With suitable corrections for losses by respiration and predation this provides a reasonable estimate of primary production. A similar method is not possible with phytoplankton because generation times are much shorter than those of higher plants and biomass is usually consumed by predators at about the same rate as it is produced. In this situation only determinations of photosynthesis made over short periods, of hours rather than days, and in the absence of predators can give an accurate idea of the rate of primary production.

2.3.2 NET AND GROSS PHOTOSYNTHESIS

All determinations of photosynthesis are fraught with uncertainty about the proper allowance for the respiration which inevitably goes on at the same time. Ideally we are concerned with the total input of potential chemical energy, i.e. with *gross photosynthesis*, but part of this total is immediately used by the plant in respiration, which, in its overall effects, is the exact reverse of the reaction given in Equation 2.i). Often, *net photosynthesis*, which is the excess of photosynthetic production remaining after respiration has taken place, is the more easily measured, and, for ecological purposes, the more useful as it is the amount available for consumption by herbivores. Conventionally, gross photosynthesis is obtained from this by adding an amount equivalent to the respiration which has occurred in similar samples kept dark but this is not really satisfactory because rates of respiration may be altered by illumination and plants show an additional light-stimulated form of respiration, *photorespiration*. The position is even more complicated when one is dealing with samples of natural populations rather than pure cultures of algae because here one has, in addition, bacteria, fungi and protozoa and other microscopic zooplankton which are not photosynthetic but which are, of course, respiring. A measurement with such a sample gives a value for the net accumulation of carbon in the community which is equal to gross photosynthesis minus the total respiration of everything, plants and animals. The more comprehensive our sample the relatively greater will the respiration be. In the limiting case where the sample is an entire ecosystem and the determination is conducted over a year, gross photosynthesis will be more or less balanced by respiration (unless accumulation of organic material is occurring, as in a peat bog) and the accumulation of carbon will be around zero.

2.3.3 METHODS

Details of techniques can be found in the manual edited by Vollenweider (1974). Here it will suffice to mention that there are two principal methods—the oxygen method, which is the less sensitive but which on theoretical grounds is the more reliable, and the radiocarbon method, which can be made highly sensitive but which may produce equivocal results, particularly because radiocarbon-labelled products may be released from the cells into the surrounding water (Fogg 1975a). A third method which is sometimes used depends on the assumption that photosynthesis is directly related to the concentration of chlorophyll in the water. Since the assimilation number or ratio (mg C fixed per mg chlorophyll per hour at light saturation) of phytoplankton may vary between 0.1 and over 50 according to conditions this is not a reliable method although an average value of 3.7 for the assimilation number often yields a good approximation to results

26

obtained by other methods. All methods are subject to sampling errors and the uncertainty about the correction necessary for respiration referred to above (section 2.3.2). For these and other reasons determinations of primary productivity made by different methods are rarely comparable, a matter which is discussed in more detail by Vollenweider (1974) and Fogg (1975 b).

2.4   The organisms responsible for primary production

Usually, primary production is referred to the total phytoplankton, measured in terms of chlorophyll content, numbers or cell volume. Numbers give only a poor idea of photosynthetic capacity since the cells counted may range from flagellates only 1 μm or so in diameter to large diatoms, such as *Ethmodiscus rex* about 2 mm across. Cell volume is a better measure but cell surface is better still. Small cells have a high surface/volume ratio, which favours rapid exchange of materials between protoplasm and environment, and, as we might expect, they are most active in photosynthesis. Phytoplankton is divided rather arbitrarily into net plankton, retained by the finest bolting silk with 64 μm apertures, and nannoplankton which passes through such a net. It is usually found that, after passing through the net, lake or sea water retains 50% or sometimes more of its photosynthetic capacity. Floristic studies on phytoplankton have until recently been concentrated on net plankton but it now appears that the smaller flagellates, diatoms, green and blue-green algae which make up the nannoplankton and which have been relatively ignored by taxonomists are at least as important in primary productivity. The same conclusion was reached in a different sort of study in which radioautography was used to determine fixation of ^{14}C by individual cells of different species in phytoplankton populations. Photosynthesis by nannoplankton was found to be greater than its biomass would suggest whereas several species were found to contribute significantly to biomass but not to primary production. Many organisms in the plankton are principally holozoic (eating particulate matter) or osmotrophic (absorbing dissolved organic matter) in nutrition but retain sufficient photosynthetic capacity to suffice for maintenance. Thus, the photosynthetic capacity of different species varies but nevertheless it remains true that the primary production of the total phytoplankton in a given body of water is primarily determined by physical and chemical factors and not by the kinds of species present.

2.5   The relation of phytoplankton photosynthesis to light

2.5.1   THE PENETRATION OF LIGHT INTO WATER

As it penetrates through water, light becomes attenuated and altered in spectral composition (Steemann Nielson 1975). The attenuation in irradiance* determines the relation of rate of photosynthesis to depth. The change in spectral

---

*When dealing with photosynthesis the rate of supply of radiant energy, irradiance, should be expressed in terms of quanta per unit area per second or of energy per unit area (1 watt cm^{-2} = 14.3 Ly min^{-1}; Ly = Langley; 1 Ly = 1 cal cm^{-2}). Lux and foot-candles, which are still often used, are units of illuminance, that is, of intensity as perceived by the human eye and have little meaning in studies of aquatic photosynthesis.

composition, which is from predominantly yellow light at the surface to blue-green at depth in clear water or yellow-green in waters containing high concentration of dissolved organic matter, is of little importance for present purposes since all wavelengths in the visible spectrum are used to some extent in photosynthesis. It is worth noting in passing however, that evidence is coming forward to show that the chemical nature of the products of photosynthesis and the development and behaviour of plankton algae may be affected by the specific wavelengths reaching the cells.

As it passes through a uniform body of water, light is absorbed according to the Lambert-Beer Law, irradiance decreasing exponentially (Fig. 2.1b) according to the expression:

$$I_z = I_0 e^{-\eta z} \qquad\qquad\qquad (Equation\ 2.iii)$$

where I_0 and I_z are the irradiances immediately below the surface and at depth z respectively and η is a constant for any given wavelength (the extinction coefficient). A natural water can be looked on as a three-component filter with absorption by the water itself, absorption by coloured dissolved materials, e.g. humic acids, and scattering by suspended matter (including the plankton as well as detritus) all contributing to the extinction.

2.5.2   VARIATION OF PHOTOSYNTHESIS WITH DEPTH

The relation of rate of photosynthesis of an alga to irradiance as determined under laboratory conditions is as shown in Fig. 2.1.a. Photosynthesis does not manifest itself until a certain threshold irradiance is passed at which the gas exchanges which it brings about exactly compensate those produced by respiration. Thereafter, light being limiting, the rate is proportional to irradiance until this reaches—very roughly—a twentieth of full sunlight, when light saturation occurs and the rate becomes independent of irradiance. At higher irradiances, perhaps around half full sunlight, photosynthesis usually becomes inhibited and the rate falls off. Combining this curve with that for light penetration into water we can predict how phytoplankton photosynthesis should be related to depth (Fig. 2.1c). At the surface, assuming a sunny day, there is strong inhibition and rate of photosynthesis increases as one goes deeper, to reach a maximum which in moderately clear water may be at 2 or 3 m depth. Below this maximum, photosynthesis falls off proportionately to irradiance. The compensation point, which is usually, but not always justifiably, taken as the depth to which 1% of surface light penetrates, may be as far down as 100 m in clear water but in turbid water it may be at only a few centimetres' depth. The layer within which photosynthesis is possible is termed the *photic* (or *euphotic*) zone; it cannot be defined precisely since its lower limit continually shifts according to time of day and other conditions. Measurements of primary production conform to this pattern if the phytoplankton is uniformly distributed in the vertical direction (Rodhe 1965; Steemann Nielsen 1975). If there are population maxima at particular depths then the curve will be distorted correspondingly.

Several features of this depth profile call for comment. Firstly, given similar solar radiation its shape is essentially the same however productive or

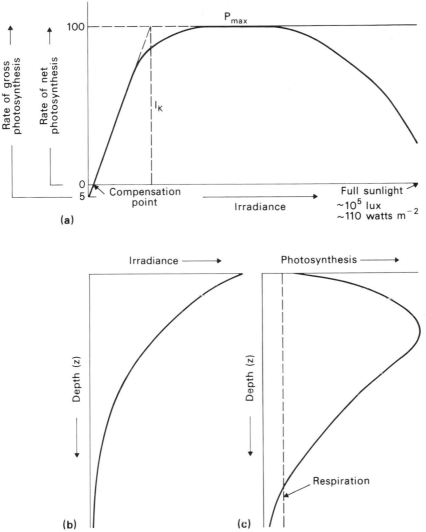

Fig. 2.1. (a) A generalized diagram showing the relationship of rate of photosynthesis by phytoplankton to irradiance. It should be especially noted that the irradiance levels at which the inflexions of the curve occur vary considerably according to species and their physiological conditions. (For explanation of P_{max} and I_k see p. 30); (b) light penetration into water with uniformly distributed plankton and other suspended matter; (c) The depth profile of photosynthesis by a uniformly distributed phytoplankton population on a sunny day as predicted from the relationships shown in (a) and (b).

unproductive, salt or fresh, the water. Ignoring the saturation and inhibition phenomena at the surface it is represented by an expression due to Talling:

$$P = \frac{nP_{max}}{k_e} \cdot \ln \frac{I_0}{0.5\, I_k} \qquad\qquad (Equation\ 2.iv)$$

in which P is the total photosynthesis beneath unit water surface, n is the density

29

of the phytoplankton population, I_0 the irradiance immediately below the water surface, I_k the irradiance at which light saturation of photosynthesis sets in (see Fig. 2.1a), k_e the vertical extinction coefficient of the photosynthetically effective radiation, and P_{max} the rate of photosynthesis per unit of population at light saturation. In clear water the maximum is several meters down and not very pronounced; below it the curve gradually approaches zero at a depth of perhaps 30 m or more. In productive water, turbid with plankton, the maximum is near the surface and very pronounced and below it photosynthesis quickly decreases to zero. Rodhe (1965) has shown that all depth profiles for photosynthesis conform to the same standard curve if photosynthetic rate is plotted as percentage of the maximum against optical depth (each unit of which corresponds to a layer of water which produces a halving of the irradiance of the penetrating light), instead of depth in metres. Empirically it is found that a reasonable approximation to the primary production below one square metre of water surface $(\sum P)$ can be obtained from a single measurement of rate of photosynthesis at the maximum ($a_{max} = nP_{max}$ in Equation 2.iv) and a measurement of light penetration, using the expression:

$$\sum P = a_{max}\, z_{0.1\ I_{mpc}} \qquad\qquad (Equation\ 2.v)$$

where $z_{0.1\ I_{mpc}}$ is the depth at which the intensity of the most deeply penetrating component of the light, usually green, is reduced to $1/10$ of its surface value.

2.5.3   SURFACE INHIBITION

The inhibition at the surface is a complex phenomenon. Sometimes it is only apparent, the lower photosynthetic rate arising simply because phytoplankton has settled out leaving a smaller population density at the surface. Also, since nutrients tend to be depleted in surface waters, the extent of extracellular release of photosynthetic products is greater (see section 2.6.4), sometimes amounting to as much as 90% of the total fixation, so that total photosynthesis may not be reduced as much as carbon fixation in the cells themselves. Mostly, however, there is a real reduction in rate of total photosynthesis per unit of population and this reduction is related principally to irradiance. The inhibition is largely due to visible light energy being absorbed at such a rate by the pigments that it cannot be used via the normal photosynthetic channels and overflows into destructive photo-oxidation reactions. Ultraviolet light also has directly destructive effects. High rates of photosynthesis tend to produce supersaturation with oxygen of surface waters and this probably increases the rate both of photo-oxidation and of photorespiration. The physiological state of the cells is important, too. Nutrient deficient cells are inhibited to a greater extent than are cells with ample nutrients under otherwise similar conditions and they also take longer to recover from inhibition (healthy cells recover in a few hours) or may even be killed.

On dull days the irradiance at the water surface may be less than saturating. In this case there will be no photoinhibition and the maximum rate of photosynthesis will be at the surface, the photosynthesis curve below this falling off smoothly with depth in parallel to that for light intensity.

Minor variations in the photosynthesis/depth curve may arise as a result of adaptation by phytoplankton. If the water column is stable so that a given cell and its progeny remain at about the same depth for a period corresponding to one or more cell division cycles then the cell characteristics, principally perhaps the pigment content, may become modified in response to the prevailing light regime (Steemann-Nielsen 1975). It is characteristic of cells grown under low light that they have a higher pigment content than those grown under high light so that under comparable conditions the 'shade' cell is able to absorb more light and thus on a per cell basis achieve a higher photosynthetic rate than the 'sun' cell. 'Sun' cells, on the other hand, are better able to withstand high irradiances than are shade cells. In summer when the water-column is thermally stratified, phytoplankton samples from near the surface show little inhibition in strong sunlight whereas those taken from a depth to which only 1% of the incident radiation penetrates are strongly inhibited by only one quarter of full sunlight, although at limiting light intensities they are photosynthetically more efficient than cells from the surface. The effect of these adaptations is to increase photosynthetic rates near the surface and in the lower part of the photic zone on bright days, but to decrease surface rates on overcast days. In late autumn, when the water column in temperate lakes and seas is completely mixed, the residence time of any given algal cell at any one depth is insufficient for adaptation to occur and samples taken at different depth show the same properties.

2.5.5 DIURNAL CHANGES IN THE PHOTOSYNTHESIS/DEPTH CURVE

Results such as just described are obtained in short term experiments, sometimes lasting over 24 hours but more usually for a few hours only, and give no idea of the way in which the pattern of photosynthetic activity in a water column continually changes with time. Obviously, at dawn or dusk the depth profile of photosynthesis will be similar to that at noon on a dull day, with the maximum at the surface. As the irradiance increases, first saturation then inhibition will

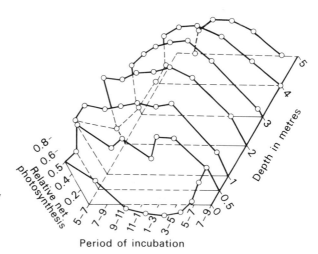

Fig. 2.2. Three-dimensional diagram showing diurnal changes in photosynthetic rate of an *Oscillatoria* population at different depths in Lake Itasca, Minnesota. [After Baker A.L., Brook A.J. & Klemer A.R. 1969. Some photosynthetic characteristics of a naturally occurring population of *Oscillatoria agardhii* Gomont. *Limnol. Oceanogr.* **14**, 327–333, Fig. 2]

Period of incubation

31

appear at the surface and the maximum will move downwards (Fig. 2.2). After noon the reverse will take place but the time-course of photosynthetic rate at any particular depth is not necessarily symmetrical about midday. It is found fairly generally that for a given phytoplankton population photosynthetic activity is less in the afternoon than it is in the morning. This is often marked in the tropics, where the rate shown in the evening may be only one tenth of that shown by a similar sample taken in the morning. Two-fold variations are usual in temperate regions, and in polar waters the effect seems to disappear. The cause of this variation is not altogether clear. In the course of the day nutrients are used up, waste products accumulate and photoinhibition effects mount up. These may all contribute but perhaps another cause is that under natural conditions cell divisions become synchronized to some extent and usually take place at night. As a result morning populations have a preponderance of young, photosynthetically active cells whereas afternoon populations have more mature, less photosynthetically active cells. This fits in with the variation with latitude since the continuous daylight at high latitudes would not be expected to induce synchrony whereas the 12 hr light:12 hr dark of the tropics would.

2.6   Primary production and nutrients

Photosynthesis and growth are interdependent processes. Photosynthesis may proceed for some time with only water and carbon dioxide as raw materials but growth is dependent on the availability of various mineral nutrients (Fogg 1975c; Stewart 1974). If these are withheld not only is growth prevented so that the photosynthetic capacity of the population ceases to expand but the photosynthetic activity of the individual cell declines as its metabolism adjusts to the deficiency. Some knowledge of the mineral nutrition of phytoplankton is therefore necessary for an understanding of primary production.

2.6.1   THE NUTRIENT REQUIREMENTS OF PHYTOPLANKTON

The mineral element requirements of plankton algae are much the same as those of other plants: in addition to the carbon, hydrogen and oxygen used in photosynthesis, nitrogen, phosphorus, potassium, calcium, magnesium, sulphur, iron, copper, zinc, manganese, molybdenum, cobalt and chlorine are needed (Stewart 1974). Sodium is necessary for the growth of blue-green algae; its concentration often affects the growth of other kinds of algae but it is not known whether it is essential for these as well. The diatoms and, possibly, the Chrysophyceae, both of which produce wall or skeletal material containing silica, have a requirement for silicon but other algal groups do not. Boron has been shown to be essential for diatoms and it may be also for other groups. In any particular water body any one or more of these elements may be in short supply. In sea-water, for example, there are ample amounts of most of the essential elements but nitrogen, phosphorus and iron are usually present in low concentrations and are the most usual limiting factors for phytoplankton growth. The same elements are usually limiting in freshwaters but occasionally other elements, magnesium or molybdenum for instance, may be limiting.

Although it seems that mineral nutrition usually controls production it must not be forgotten that some 70% of planktonic algae that have been examined from this point of view have absolute requirements for one or more organic growth factors or vitamins (Stewart 1974). A requirement for vitamin B_{12} (cobalamin or one of its analogues) is most common but vitamin B_1 (thiamine) or vitamin H (biotin or coenzyme R) are essential for the growth of some species. It is possible that other growth factors remain to be identified since samples of natural water often seem to contain growth promoting substances which cannot be replaced by known compounds. Algae requiring organic growth factors, *auxotrophs*, may be fully photosynthetic; their deficiency lies in inability to synthesize a particular chemical grouping required for an enzyme or other essential cell constituent and the amount of organic carbon derived from the vitamin is negligible. Auxotrophs are found in most, if not all, algal groups but it is the Chrysophyceae, Haptophyceae, Cryptophyceae and Dinophyceae that have the largest proportions of such species.

2.6.2   THE UPTAKE OF NUTRIENTS

Whether it be a mineral ion or a vitamin, the physiology of uptake is much the same. It is necessary first of all that the molecule or ion should diffuse up to the cell surface. Since with limiting nutrients we are dealing with extremely low concentrations—phosphate concentrations in seawater are often as low as $3 \mu g$ P l^{-1} and vitamin B_{12} is usually present at about 1 ng l^{-1}—the water around an algal cell rapidly becomes depleted and if cell and water are stationary relative to each other a diffusion gradient is set up. The rate of uptake is then limited by the diffusion process. If, however, cell and water move relatively to each other the supply is replenished and the diffusion gradient is steepened. It is thus important for uptake of scarce nutrients that phytoplankton cells should be able to move through water. Many vitamin-requirers are flagellates and achieve this 'forced convection' by active swimming. In dinoflagellates, for example, in addition to the forward propulsion provided by the longitudinal flagellum, there is rotation, provided by the transverse flagellum, which must increase the flow of water over the cell surface (Margalef 1978). Non-motile cells can achieve the same effect by sinking and, paradoxical though it may seem, sinking is of vital importance for them. For diatoms, for example, it has been shown that division rates are related to their calculated sinking rates. Water circulation is usually sufficient to return a high proportion of sinking cells to the surface so that the uptake of nutrients and photosynthesis are not actually incompatible (see section 2.8.1).

Uptake of nutrients seems usually to be by active processes, capable of operating against a concentration gradient between cell and environment, and therefore requiring a supply of energy (Stewart 1974). Energy may be supplied either by oxidative phosphorylation in respiration or by photosynthetic phosphorylation and the dependence on one or other of these processes varies not only from species to species but during the life cycle of a cell. The marine diatom *Skeletonema costatum* takes up nitrate and ammonium nitrogen mainly during the light whereas the marine flagellate *Coccolithus huxleyi* takes up both nitrogen sources at a high rate in both light and dark. The uptake mechanisms may be extremely efficient; the diatom *Phaeodactylum tricornutum* may reduce the phosphate in seawater to as little as 0.02 μg P l^{-1}, a concentration that is scarcely measurable by the usual analytical techniques. Uptake increases proportionately to concentration at low nutrient concentrations but at higher concentrations approaches a maximum and becomes independent of concentration. The relationship is described by the expression:

$$u = \frac{u_\infty c}{c + K_c} \qquad\qquad (Equation\ 2.vi)$$

where u is the rate of uptake, u_∞ the maximum rate, c the concentration and K_c a constant which is numerically equal to the concentration giving half the maximum rate of uptake. K_c is a useful measure (inverse) of the ability of phytoplankton species to take up a given nutrient and provides an indication of ecological behaviour. Thus it is found that oceanic species have lower K_c values than coastal species accustomed to nutrient richer waters. K_c values are usually of the same order as the nutrient concentrations found in natural waters.

2.6.3 THE RELATIONSHIP OF GROWTH RATE TO NUTRIENT CONCENTRATION

If a particular nutrient is at limiting concentration then one would expect its rate of uptake to be the bottle neck, so that the relative growth rate of the organism, k, would bear the same relationship to concentration as u does in Equation 2.vi. This is often so, but there are complications (Fogg 1975c). One is that two or more nutrients may be limiting simultaneously. There is a gradual transition, not an abrupt change, from the limiting to the saturating situation, and if the concentrations of several nutrients are in this transitional range then change in any one of them may produce a change in growth rate. Another is that it may be uncertain what the effective concentration of an ion is, since the dissolved organic components of natural waters form complexes with inorganic ions and these may be more, or less, available to algae than the free ions. It should also be noted that the concentration of a nutrient is no indication of the actual supply available. Mineral nutrients are continually regenerated in a natural plankton community by excretion and bacterial decomposition so that their turnover may be rapid and the flux large. Turnover times of inorganic phosphate in temperate seas have, for example, been found to be as little as 1.5 days (Golterman 1975). Growth rate, in any case, is determined by the intracellular concentration of a nutrient and bears no direct relationship to the external concentration. Thus, if, as sometimes happens, there has been 'luxury consumption' a cell may contain a sufficient concentration to maintain a high rate of growth in spite of the concentration in the external medium being inadequate. Algal cells, in general, if grown on an ample supply of combined nitrogen will accumulate sufficient to carry them through two successive divisions in the absence of any further supply. Phosphate is stored by algal cells as polyphosphate and the freshwater diatom *Asterionella formosa* has been shown to be capable of storing in this way enough reserves to provide for nearly seven doublings. Behaviour may also play a part in complicating the issue. The

34

larger dinoflagellates may show daily migrations over a vertical range of 10 m or so and blue-green algae, by means of their gas-vacuole buoyancy regulating mechanism, may also move up and down in the water column over smaller distances (Fogg 1975c). Since nutrients are usually most depleted near the surface there is the possibility that such species may be able to replenish their nutrient supplies in deeper water after photosynthesizing nearer the surface. Because of these various complications a correlation between concentration of a nutrient in a natural water at a particular time and phytoplankton growth and photosynthesis at the same time is not to be expected and is not usually found. Indeed, there is most often an inverse relationship; thus it is generally observed that blooms of planktonic blue-green algae appear in freshwaters when concentrations of phosphate and nitrate are at their lowest. At such times addition of nutrients may stimulate photosynthesis and this stimulation, as measured by the radiocarbon technique, provides a sensitive means of detecting which nutrients are limiting.

2.6.4   THE RELATIONSHIP OF PRIMARY PRODUCTIVITY TO NUTRIENT
         SUPPLY

Nevertheless there is a general correlation between the nutrient status of a body of water and its primary productivity (see pp. 16–19). At the extremes, nutrient-poor, oligotrophic waters such as the Sargasso Sea and alpine lakes may fix only $0.2 \mathrm{~g~C~m}^{-2} \mathrm{~day}^{-1}$ whereas nutrient-rich, eutrophic, waters such as those off the coast of Peru or shallow lakes receiving drainage from agricultural land may sometimes fix as much as $10 \mathrm{~g~C~m}^{-2} \mathrm{~day}^{-1}$. There is a continuous gradation between oligotrophic and eutrophic and a hard and fast distinction between them is not possible. Nevertheless the situations which give rise to the two conditions are usually clear. Most of the oceans are oligotrophic (Fig. 2.3), the major anticyclonic circulations north and south of the equator in the Atlantic, Pacific and Indian Oceans causing water to sink in the central parts and draw in nutrient-deficient surface water centripetally from the margins. In the tropics, oligotrophic conditions are perpetuated by thermal stratification which stabilizes the water column and minimizes mixing of cold bottom water, charged with nutrients by mineralization of sedimented organic material, into the warm depleted photic zone above it (Cushing 1975). Eutrophic conditions occur where nutrient rich bottom water upwells to the surface. This happens on the western coasts of continents, where off-shore winds blow surface water away from the shore, and where currents diverge (Fig. 2.3). Upwelling in the Indian Ocean is related to the monsoons. The rich surface waters of the Southern Ocean, which have from three to 40 times higher concentrations of plant nutrients than Arctic waters, are derived from deep water flowing from equatorial regions which upwells to replace water sinking as it is cooled in the vicinity of the Antarctic continent. Inshore situations, which are usually nutrient-rich, are more eutrophic than the open sea. Oligotrophic lakes are characteristically found in hard-rock regions, the drainage from which brings in little dissolved mineral matter. These lakes are usually deep so that the greater proportion of the available nutrients are locked up in the cold bottom water (*hypolimnion*) and unavailable to the phytoplankton in the warm upper layer (*epilimnion*) during summer stratifica-

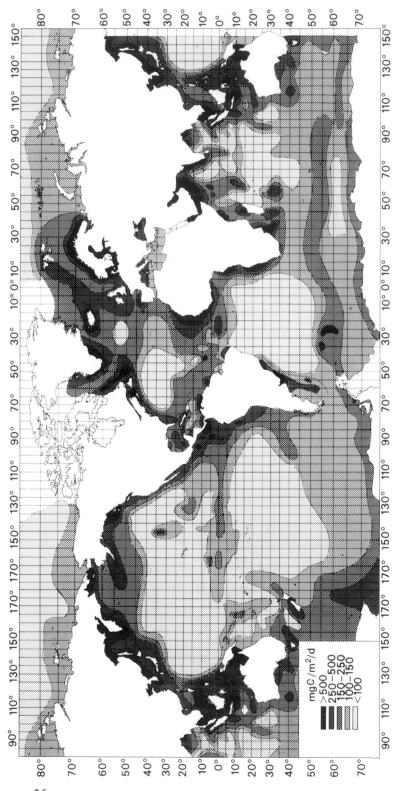

Fig. 2.3. Primary production in the world's oceans in mg C fixed per square metre per day. [After map 1.1. in *Atlas of the Living Resources of the Seas* F.A.O., Rome 1972]

mgC/m²/d

>500
250–500
150–250
100–150
<100

tion. Eutrophic lakes are characteristically found in lowland regions and usually their catchment areas are under cultivation, so that ample nutrients are available from both natural and artificial sources. They are usually shallow, with a high epilimnion/hypolimnion ratio and, since the small amount of oxygen held in the hypolimnion is used up by decomposition processes during summer, the bottom water may become anaerobic and this promotes release of nutrients from bottom deposits. In the autumn, when thermal stratification breaks down in temperate lakes these nutrients become mixed into the general body of water. Sometimes lakes are permanently stratified, as for example in tropical regions or in temperate regions if a density gradient, produced by accumulation of dissolved material in the hypolimnion, augments the stabilizing influence of a temperature gradient. Such lakes are unproductive since the epilimnion is continually depleted by organic remains sedimenting out and diffusion unaided by mixing is insufficient to return any appreciable amount of nutrient from the hypolimnion.

Inland seas, coastal lagoons and estuaries, generally show high productivity because in temperate regions at least, they receive a high proportion of river water and runoff from the land which promotes mixing of the coastal waters and brings in nutrients such as nitrate and phosphate as well as organic detritus. The concentrations of dissolved substances in their waters are not entirely a simple resultant of the proportions of fresh and salt water which have been brought together. Adsorption of dissolved substances on flocculated suspended matter and interaction with sediments may considerably modify the chemistry. Although nutrients may be present in high concentrations in estuaries the phytoplankton cannot attain maximum productivity because light penetration is restricted by the turbidity of the water. The phytoplankton is dominated by marine species but because of fluctuations in salinity oceanic species are unable to persist. Blooms of coastal species may invade an estuary or occur in the dominantly marine part but there are also characteristic estuarine species which bloom in the upper or middle sections and extend downstream. Such species are the diatoms *Asterionella japonica* and *Nitzschia seriata* and dinoflagellates such as *Peridinium ovatum*. Lagoons in the tropics and subtropics may become highly saline through evaporation. This does not necessarily limit productivity (see Table 1.1) and high biomasses of planktonic species such as *Dunaliella salina* may occur.

An effect of availability of nutrients on the release of extracellular products of phytoplankton photosynthesis may be noted here. This release is a normal occurrence and glycollic acid, an early product of photosynthesis, is usually prominent (Fogg 1975a). The absolute amount of organically combined carbon which is lost from the cells is perhaps much the same whatever the type of water but the amount relative to the total fixation varies from less than 1% in eutrophic waters to 40% or more in oligotrophic waters. The release has the characteristics of an overflow process in that anything, such as nutrient deficiency, which limits the capacity of cells to grow and take up photosynthetic products as fast as they are formed, tends to increase the extent of the release. This release of extracellular products is important quantitatively since un-productive waters predominate in the oceans and estimates of primary productivity which ignore it are consequently serious underestimates.

2.7   Turbulence and primary production

2.7.1   TURBULENCE AS A LIMITING FACTOR

A further most important factor influencing primary production is turbulent mixing. Light is available only near the water surface whereas there is a continual tendency for nutrient elements to accumulate at the bottom as a result of sedimentation. Except to a limited extent by vertical migration, phytoplankton has no means of overcoming the resulting spatial separation of these two essentials, and as we have seen, if the water remains stagnant the end result is an aquatic desert with a minimum of phytoplankton. Nutrients are returned to the surface and phytoplankton production becomes possible as a result of upwelling or turbulence and this depends on energy input into the water body by solar radiation, wind and tidal action. There is consequently a direct relationship between primary production by phytoplankton and the input of external energy into the ecosystem. This energy factor overrides light and nutrient concentrations as a determinant of primary production in open seas and lakes (Margalef 1978).

There is, however, an upper limit to the beneficial effects of turbulence. If the sea or a lake is mixed, say by wind action, then a cell will be carried between the surface and depth and the light available to it for photosynthesis will be the average over the water column which it traverses. If mixing is vigorous the cell will go deeper and get less light on average so that effectively it will be below the light compensation point and growth will not be possible (unless heterotrophic growth at the expense of dissolved organic substances takes place; this may be possible but there is no direct evidence that it occurs in the majority of phytoplankton species). A critical depth can be defined at which the photosynthesis in the water column above is equal to the respiration. If the depth of turbulent mixing exceeds this critical depth then the phytoplankton population cannot grow. In winter in temperate waters there are ample nutrients, the low temperature is not altogether restricting, there is sufficient light for some algal growth and, indeed, in sheltered waters phytoplankton growth does occur. Phytoplankton is sometimes remarkably abundant and active under ice. Mostly, however, wind induced turbulence in the sea exceeds the critical depth in the winter and phytoplankton growth does not occur. In lakes this may also be so but washing out of phytoplankton by increased flow of water through the lake may be important in preventing growth too. Artificial circulation of reservoirs has been found a useful method of controlling phytoplankton growth in drinking water supplies and this evidently depends, among other things, on this same effect. As mentioned above the concentration of nutrients in the Southern Ocean is exceptionally high but although light is ample in the austral summer, phytoplankton production is minimal over most of it because of turbulence induced by the prevalent high winds. Rivers tend to be turbid, the critical depth is therefore small, as well as turbulent, so that mixing is deep, and thus they do not often support phytoplankton growth even if the rate of flow would otherwise permit this.

As winter turns to spring in temperate regions irradiance increases, day length increases and winds moderate so there comes a point where the depth of mixing is less than the critical depth and growth begins (Fig. 2.4), often so

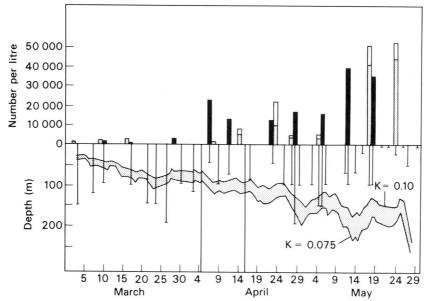

Fig. 2.4. Depth of the superficial mixed layer and phytoplankton growth: Observations at Weather Ship M (66°N, 2°D) from March to April 1949. Upper solid bars, phytoplankton cells per litre; vertical lines represent the depth of the superficial mixed layer (Dm) and the shaded zone represents the critical depth (Dcr) calculated as lying between two values of the extinction coefficient for light (k = 0.10 or 0.075). Growth of phytoplankton only occurs when Dm is equal to or less than Dcr. [After Sverdrup H.U. 1953. On conditions for the vernal blooming of phytoplankton. *J. Cons.* **18** 287–295]

suddenly that it merits the name of 'outburst' or 'bloom'. Inflow of fresh water from rivers may also stabilize the water column and allow phytoplankton growth and, in the Antarctic, freshwater from melting ice may produce the same effect. In shallow lakes and ponds the depth of the water may nearly always be less than the critical depth. In this situation turbulence, by ensuring rapid interchange of materials between water and mud and by removing cells rapidly from inhibitory irradiance near the surface, promotes phytoplankton production.

2.7.2 PRIMARY PRODUCTION AT FRONTS

A special manifestation of mixing found in seas and large lakes is the front, which is essentially a sharp boundary between water masses of different properties. Fronts may arise from various causes but nearly always have increased biological activity associated with them. As an example, fronts produced in shallow seas by tidal mixing may be discussed. The mixing effect of the tide depends on its velocity and the depth of the water and is opposed by the stabilizing effect of thermal stratification produced by solar radiation. Hence in shallow waters with a shelving bottom there may be, at a particular depth, a transition between an area in which tidal mixing of the entire water column occurs and one which is thermally stratified. From a ship this may be detected by a drop in surface temperature as it passes across the front from the stratified to the mixed area. Fronts also show up quite distinctly in infra-red photographs taken from satellites, because of this temperature difference. There are many around Britain and those in the English Channel and the Irish Sea have been studied in particular (Pingree *et al.* 1975).

Besides being marked by the temperature difference, a front is also usually distinguished by having a high phytoplankton population associated with it. This may be explained, at least partly, in the terms which have just been discussed. On the mixed side of the front nutrients are in ample supply because bottom water is continually brought back to the surface but, although the critical depth may not actually be exceeded, the effective irradiance to which the phytoplankton population is exposed is reduced so that growth is poor. On the stratified side the phytoplankton is able to remain in the photic zone but this is cut off by the thermocline from the nutrient rich bottom water, so, again, growth is poor. It is interesting that on the mixed side of the front the dominant organisms tend to be diatoms (which are dependent on turbulence)

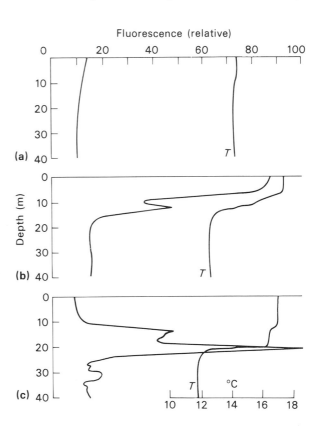

Fig. 2.5. Vertical profiles of chlorophyll fluorescence and temperature in the vicinity of a front off Ushant in the Southwestern Approaches to the English Channel in July 1975. (a) well mixed; (b) frontal and (c) stratified regions. [After Pingree R.D., Pugh P.R., Holligan P.M. & Forster G.R. 1975. Summer phytoplankton blooms and red tides along tidal fronts on the approaches to the English Channel. *Nature* **258**, 672–7, Fig. 4]

whereas on the stratified side motile dinoflagellates predominate. At the front itself there is both sufficient mixing to bring up nutrients and sufficient stability for the plankton to remain in the photic zone and growth is correspondingly greater (Fig. 2.5). There may be additional factors; for example the water on one side may be deficient in one particular nutrient whilst a different one may be limiting on the other side so that at the front the two waters complement each other and produce a mixture better able to support growth.

2.8   Spatial variation

2.8.1   PATCHINESS

A lake or sea may give the impression of being well mixed and uniform but the phytoplankton is rarely distributed evenly. Uneven distribution is obvious when, as a result of strong winds, scums of buoyant species such as *Botryococcus*, flagellates or blue-green algae accumulate to leeward on a lake. From the air under suitable conditions plankton can often be seen distributed in patches, streaks or convoluted bands, varying from a metre or so across up to one or two hundred kilometres. The streaks of phytoplankton some few metres apart which often develop in light and moderate winds are associated with Langmuir circulation. Rotating 'cells' of water with their long axes parallel to or at a slight angle to its direction are set up by the wind. Where there is downwelling between two of these 'cells' buoyant phytoplankton or floating material will be drawn into a line on the water surface at the convergence. Phytoplankton cells with a tendency to sink will have their descent accelerated in this area. On the other hand in the upwelling parts of the 'cell' the rate of movement of the water upwards may be greater than the rate of sinking of phytoplankton so that the population is maintained near the surface, again in parallel lines. In this and other ways, populations of dinoflagellates such as *Goniaulax* spp. may be concentrated to several million cells per litre from populations only a hundredth or less of this. It will be realized that Langmuir circulation also provides a means of segregating species. Whether a concentration once formed is dispersed or not will depend on the ratio of rate of mixing to the rate of reproduction of the species. In water which is thermally stratified and subject only to slight wind-induced turbulence there is a distinct tendency for different species to be present in separate 'clouds'.

To get an accurate idea of the mean primary productivity of a water body in which plankton is distributed patchily requires that a large number of stations should be sampled. This is rarely possible and it has to be admitted that most estimates of the primary productivity of large lakes or sea areas, being based on a few unrepresentative determinations, can only be regarded as rough approximations.

2.8.2   SPECIES DISTRIBUTION

The large scale variations of phytoplankton productivity which are summarized in Fig. 2.3, are generally explicable in the terms discussed in section 2.6.4. There are, of course, differences in the distribution of species as well as in primary productivity. Different algal groups tend to dominate the phytoplankton floras of sea and freshwater. The main marine groups are the diatoms (Bacillariophyceae), Dinophyceae, Chrysophyceae and Haptophyceae, with the blue-green algae (Cyanobacteria) abundant only in certain tropical areas. Dinophyceae are sporadic in Antaric waters, the flora of which is dominated by diatoms, but are extremely abundant in the tropics. This may indicate different temperature preferences but it may also be another result of the diatom's requirement for turbulence and the dinoflagellate's ability to swim.

In freshwaters the diatoms and Dinophyceae are again abundant but the Chrysophyceae and Haptophyceae are less prominent and the green algae (Chlorophyceae) and blue-green algae are often dominant. Freshwater plankton species have a distinct tendency to be cosmopolitan in distribution.

The diversity index of a phytoplankton sample tends to be lower in eutrophic waters than in oligotrophic waters. Also, species with r-strategy tend to predominate in eutrophic waters whereas the K-strategy characterizes the plankton of oligotrophic waters. It may be noted here that the diversity of phytoplankton samples gives rise to a problem since it is difficult to see how so many species with similar requirements could continue to coexist in an apparently uniform environment. This has been called by Hutchinson 'the paradox of the plankton'. The explanation probably is that the uniformity is only apparent and the continually changing pattern of turbulence offers a succession of niches which can be occupied by a particular species for only a limited time. The survival of a population is the result of a temporary equilibrium between success in remaining afloat and inevitable sinking and different species differ in their sedimentation properties and in their capacities to take advantage of different levels of turbulence (Margalef 1978).

2.9   Seasonal variation

2.9.1   SEASONAL VARIATION IN BIOMASS

Phytoplankton growth and primary production show characteristic seasonal variations (Fogg 1975c; Goltermann 1975). Solar radiation is the prime factor concerned and, if nothing else were involved, primary production would follow the same bell-shaped curve during the year as it does; a tall bell with its edges at spring and autumn in polar regions and a shallow bell covering the whole year in the tropics. However, as we have seen, turbulence in the winter reduces primary production below what would be expected from the irradiance incident on the water surface. Once mixing diminishes to less than the critical depth, phytoplankton growth begins and the population increases rapidly by about one-thousand-fold in temperate and polar waters, somewhat later in the latter because of lower temperatures and greater turbulence (Fig. 2.6). Zooplankton populations are low during winter and these organisms are more dependent on higher temperatures for growth and development than are algae so at first the grazing pressure on the phytoplankton is slight. Soon, however, with ample food and warmer water the zooplankton populations build up, more slowly in polar waters than in temperate climes, and phytoplankton populations are reduced almost as rapidly as they increased. Exhaustion of nutrients from the photic zone also contributes to a greater or lesser extent to this decline. By the time this has happened in polar regions winter is setting in again but in temperate regions after reaching a low during the summer the phytoplankton usually is able to reach a second peak in the autumn when zooplankton populations have fallen, nutrients are becoming available again as equinoctial gales mix in bottom water, and light and temperature are still adequate. In the tropics the high temperatures enable the zooplankton to grow rapidly so that phytoplankton has little chance to build up. The oscillations in the population curves are damped

42

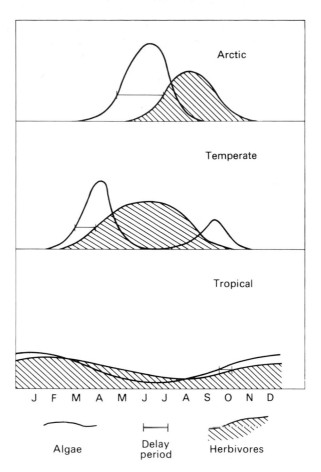

Fig. 2.6. Seasonal variations in phytoplankton and herbivorous zooplankton (hatched area) in different latitudes. The horizontal bar indicates the delay period between the increases of phytoplankton and zooplankton. [After Cushing D.H. 1959. The seasonal variation in oceanic production as a problem in population dynamics. *J. Cons.* **24**, 455–464, Fig. 3]

Arctic

Temperate

Tropical

J F M A M J J A S O N D

Algae   Delay period   Herbivores

and there is little more than a five-fold difference between maxima and minima.

These patterns are generally followed both in the sea and in freshwater but there is some variation in both causes and effects. Oceanic sea areas in temperate regions tend to conform to the polar pattern rather than the temperate pattern just described because greater turbulence makes for a later start to the cycle and temperatures are lower. In freshwaters it seems that nutrient depletion is a more important factor than grazing in producing the summer minimum and if the supply of nutrients is particularly ample, as in waters polluted with organic waste or receiving drainage from rich agricultural land (*eutrophicated* waters), there is no summer minimum and the population curve for the phyto-plankton follows the simple bell-shaped curve. In the tropics population maxima may occur in the rainy season when run-off from the land brings in nutrients. In the oligotrophic Sargasso Sea, phytoplankton numbers remain at a low and fairly constant level throughout the year except when winter temperatures fall low enough for a thermocline at about 100 m to break down and release into the photic zone small amounts of nutrients (including vitamin B_{12}, which here is limiting for diatom growth) sufficient to produce a distinct maximum in spring. Where algal production in the sea is dependent on upwelling of nutrient-rich water as in the Benguela and Peru currents, then the seasonal cycle is dominated by the cycle of upwelling.

2.9.2 SEASONAL VARIATION IN PRIMARY PRODUCTION

Generally speaking primary production is related to biomass, and seasonal cycles of primary production are of similar form to those for phytoplankton populations which have just been described. However, there is no simple general rule relating photosynthesis to biomass (as measured for example by chlorophyll; see section 2.3.3). Measurements of primary production show great day-to-day variations ascribable to changes in weather and circulation patterns whereas biomass shows much less variation. The physiological state of the algae is also important. During a phase of rapid growth such as occurs in temperate lakes in the spring there may be for a time an almost constant relationship between primary production and biomass but later on this breaks down and the ratio of the two may show variation of up to two orders of magnitude.

2.9.3 SEASONAL SUCCESSION AND PRIMARY PRODUCTION

While it is convenient for many purposes to think of the phytoplankton in terms of total biomass it must not be forgotten that this total is made up of different species and that the composition of the phytoplankton shows great variation with time (Fogg 1975c). In any given sea area or lake there is a seasonal succession which, although it may show great variation in detail, usually conforms to a general pattern. In a temperate eutrophic lake, for example, the spring maximum is usually made up largely of diatoms and these are succeeded by green algae and flagellates. Later in the summer dinoflagellates become prominent and blue-green algae are characteristic of late summer and autumn. Most studies of succession have been concerned with the easily identifiable net plankton. The few studies made on nannoplankton suggest that it may be more constant in biomass and perhaps in species composition over the year. In the sea, too, diatoms are usually the most abundant forms in spring whereas dino-flagellates are more characteristic of more stable summer conditions. The phytoplankton of the spring bloom consists of species characteristically having an r-strategy, having small size with a preference for high concentrations of nutrients and a relatively high rate of growth. These species are easily grown in crude culture. The species which prevail in summer, on the other hand, are slower growing and have complex nutritional requirements so that they are extremely difficult to culture. These characteristically have the K-strategy. This is in accord with the idea that at the start of the growing season a natural water is charged with mineral nutrients and as the season progresses these are depleted and soluble organic substances produced by excretion and the decay of organisms accumulate making it a much more complex medium for growth. Several studies, both in the sea and in lakes have shown that diversity indices for the phytoplankton are low in the spring and high in the summer and there are some indications that primary productivity is inversely correlated with diversity index.

44

2.10 The products of primary production

The biomass produced by primary production, then, consists of a variety of different species. There can be little doubt that in their grazing many herbivores select particular species according to criteria of size, shape and, presumably, taste. Taste will depend largely on secondary products peculiar to each species. On the whole the different groups and species of planktonic algae tend to resemble each other in the relative amounts of crude protein, polysaccharides and fats which they contain. Some groups, of course, are characterized by large amounts of inert skeletal or wall material, such as silica or calcium carbonate. The gross chemical composition of a species of algae may vary within wide limits according to the conditions of growth (Stewart 1974; Fogg 1975c); actively growing *Chlorella pyrenoidosa*, for example, may contain 50% protein, 32% carbohydrate and 18% fat on a dry weight basis, whereas after prolonged nitrogen starvation it contains 8.7% protein, 5.7% carbohydrate and 85.6% fat. However, the species present in any given plankton sample will have grown under approximately similar conditions and thus will be in much the same physiological state and have much the same overall chemical composition. The simple concept of the food web in the planktonic ecosystem assumes that herbivores graze directly on the phytoplankton but in view of the extent of release of extracellular products mentioned earlier in this chapter this may require modification. This release appears to be normal in both seas and lakes, is irrespective of the types of phytoplankton, and in oligotrophic waters accounts for a large proportion of the primary productivity. Glycollic acid, one of the substances released, is a normal constituent of seawater, in which it may occur in concentrations of up to $0.1 \, \text{mg} \, l^{-1}$ or more (Fogg 1975a). Glycollate can act as a substrate for bacterial growth and in both lakes and the sea glycollate-utilizing bacterial floras have been found which differ in distribution from other heterotrophs. Bacteria which have grown at the expense of glycollate and other substances of similar origin may provide a food source of appreciable importance for both planktonic and benthic animals.

Aquatic animals are capable of taking up dissolved organic compounds from the water in which they live and it is possible that to some extent they may benefit directly from release of extracellular products of photosynthesis. It is sometimes found both in the sea and in freshwater that animal populations are larger than can be accounted for in terms of the plant growth which is apparently available for them to eat and some alternative source of food such as that provided by extracellular products or bacteria seems to be necessary.

3

Zooplanktonic Production

T.R. PARSONS

3.1   Introduction

3.1.1   THE ZOOPLANKTON

Zooplankton is a general term for pelagic animals which are unable to maintain their position by swimming against the physical movement of water (p. 6). While the term 'zooplankton' is convenient ecologically as an approximate grouping of a very wide range of organisms, its precise definition is difficult. This becomes even more apparent when one considers the trophic structure of aquatic food chains. Secondary production in the pelagic environment, for example, will often be interpreted as being production by zooplankton (as opposed to production by phytoplankton, i.e. primary production). However, this is not the case since the zooplankton community contains both herbivores (e.g. many species of copepods) and carnivores (e.g. jellyfish), the latter group belonging to the tertiary producers, or even to some higher level of production (e.g. when large jellyfish feed on nekton).

Among the marine zooplankton the principal phyla are the Protozoa, Coelenterata, Ctenophora, Chaetognatha, Annelida, Mollusca, Arthropoda (class Crustacea) and the (subphylum) Urochordata. In general the ctenophores, coelenterates, chaetognaths and annelids are exclusively carnivorous and vary in size from a few millimetres to some jellyfish with tentacles of several metres. The largest group of zooplanktonic organisms are the Crustacea and these may be herbivorous, omnivorous or carnivorous, varying in size from the smallest nauplius of several micrometres diameter to large pelagic euphausids of several centimetres length. Freshwater zooplankton are generally smaller in size (p. 6) and are represented by fewer animal phyla than their marine counterparts. The principal freshwater zooplankton are various Protozoa, Rotifera, Crustacea (especially the Cladocera, Copepoda and Ostracoda) and meroplanktonic organisms, including insect larvae (e.g. the midge, *Chaoborus*).

The life cycles of the zooplankton are extremely varied and for specific groups of organisms a text on invertebrate zoology should be consulted. For the purpose of studying pelagic ecosystems it is possible to summarize the life cycles of a number of representative organisms in terms of various strategies which they employ for survival (and see pp. 162–184). In this sense a 'strategy' can be interpreted as a step in the life cycle of an organism which, over evolutionary time, has been found to be expedient for an organism's survival. For example, the migration of certain zooplankton into deep water during part of their life cycle (ontogenetic migration), in order to avoid higher predation rates in the euphotic zone, could be considered a strategy for the survival of a particular species.

A summary of some steps in the life cycles of various representative zooplankton are shown in Table 3.1. In this table it can be seen that the form of reproduction among zooplankton can vary from differentiation of adults into male and female to asexual division of polyps among coelenterates and hermaphroditic reproduction among chaetognaths. The eggs of various zooplankton become part of the plankton community if they are released by oviparous adults; in other cases, such as among amphipods and salps, the young are released directly into the water column (ovoviviparous). Metamorphosis is a common characteristic of crustaceans; for example, copepods pass through six naupliar and six copepodite stages before becoming adults. The number of generations per year may be determined in part by the water temperature and in part by the availability of food. Thus some copepods may have three to five generations per year in warm waters while the same or similar species may have only one generation per year in colder waters. In other cases a species such as the Pacific Ocean copepod, *Calanus plumchrus*, is always confined to a single generation per year since it requires a full year to complete its life cycle. Other zooplankters may require several years in which to complete their life cycle (e.g. the antarctic krill, *Euphausia superba*).

Zooplankton living their entire life cycle in the pelagic environment (holoplankton) tend to dominate in open waters, but in coastal regions there will be larger numbers of organisms which spend only part of their life cycle as plankton (meroplankton) and these will include the larvae of benthic invertebrates as well as the larvae of some benthic and pelagic fishes.

3.1.2 DISTRIBUTIONS

The distribution of zooplankton shows changes in abundance both spatially and temporally. The identification of these spatial and temporal differences in abundance is convenient in describing the zooplankton community but in fact they are interrelated and thus should not be separated when it comes to a discussion of the dynamic processes governing their occurrence. For example, it can be observed that many zooplankton migrate vertically over a 24-hour period with a maximum concentration occurring near the surface at night but at depth during the day (see pp. 49 & 165–6).

Another problem inherent in observations on distributions is that both the time and distance scales of the observations must be taken into account in reporting an observed distribution. Failure to do this leads to 'aliasing' of data. (As a simple example of aliasing, an event which might occur every 12 hours can be made to have an apparent occurrence every 36 hours if observations are made at 18-hour intervals.)

The uneven distribution of zooplankton both in lakes and seas has been apparent almost from the time when zooplankton were first collected for scientific study. It has only been recently, however, that scientific investigations have been carried beyond a purely statistical account of the non-random nature

Table 3.1. Some representative life cycle strategies of organisms to be found among the plankton.

Organisms	Common names	Reproduction	Eggs	Stages	Generations per year	Planktonic existence	Comments
Crustaceans: *Euphausia pacifica*	Euphausiid	Separate sexes	Oviparous	6	0.5 to 1	Holoplanktonic	
Pseudocalanus minutus	Copepod	Separate sexes	Oviparous	12	Ephemeral	Holoplanktonic	
Calanus plumchrus	Copepod	Separate sexes	Oviparous	12	1	Holoplanktonic	
Anisogammarus pugettensis	Amphipod	Separate sexes	Ovoviviparous	None	Ephemeral	Bentho-pelagic	
Chaetognaths: *Sagitta setosa*	Arrow worm	Hermaphroditic	Oviparous	None	0.5 to 3	Holoplanktonic	
Coelenterates: *Aurelia*	Jellyfish	Asexual and sexual (metagenesis)	Oviparous	None	1	Meroplanktonic	Alternate medusoid and hydroid forms
Molluscs: *Spiratella retroversa*	Sea snail	Hermaphroditic	Oviparous	1	Ephemeral	Holoplanktonic	
Urochordates: *Thalia democratica*	Salp	Asexual and sexual (metagenesis)	Ovoviviparous	None	Ephemeral	Holoplanktonic	
Teleosts: *Clupea harengus*	Fish (herring)	Separate sexes	Oviparous	1	0.15	Meroplanktonic	Only the larvae are planktonic

of these distributions. Some general processes leading to aggregation of zooplankton into small and large patches can be summarized as being due to one or more of the following: the presence of a physicochemical boundary, the action of wind and water circulation, reproductive strategies and intraspecific feeding and predation.

Physicochemical boundaries are particularly important in accounting for the vertical distribution of zooplankton since there are generally strong vertical gradients in terms of light, temperature and (in the sea) salinity within a relatively short distance from the surface to the bottom of any water mass. Thus it is common to observe a greater abundance of zooplankton at depth during the day both in lakes and seas, and to find that much of this abundance migrates to the surface at night. The phenomenon is particularly well marked where a large accumulation of animals (plankton and nekton) can be detected by sonar in the ocean. This accumulation is often referred to as the 'deep scattering layer' (or DSL) and it can be observed to follow an isolume to the surface as sunlight fades in the evening. The actual distance that some planktonic organisms migrate during this diel change in light intensity can be as much as 500 m, although in many cases the smaller plankton migrate from much shallower depths of 10 m or less, depending on water clarity. Some planktonic species do not exhibit diel migration and in such cases they may occupy relatively discrete zones at depth, or in the surface layers. For example, arctic plankton under conditions of continuous daylight show almost no change in their vertical abundance with time. However, vertical migration may also occur with season, such as among many species of copepods in temperate latitudes. For example, a dominant species of copepod in the temperate North Atlantic, *Calanus finmarchicus*, overwinters as a Stage V copepodite in deep water (down to 1000 m) where it eventually matures and migrates back into the surface waters during the spring. In experiments with the freshwater cladoceran *Daphnia magna*, it has been observed that aggregation of this organism is sensitive to the colour of light, and hence to the amount of particulate material in the water (the latter affecting colour through the scattering of shorter wavelengths). Strong salinity and temperature gradients may also form a boundary against which organisms will accumulate. For example, the marine copepod *Acartia bifilosa* has been found to be prevented from migrating through a salinity gradient from 34‰ to 24‰; in other cases greater sensitivities down to differences in salt content of 1‰ have been observed. Another gradient of importance in determining plankton distribution is the oxygen content; plankton are generally reported to avoid very poorly oxygenated water and to be absent from water containing hydrogen sulphide.

While vertical gradients of plankton abundance may occur over relatively short distances of 1–10 m, there are also horizontal patches to be considered. Many species may be found over thousands of kilometres (i.e. panoceanic). On the other hand, patches of individuals may also occur on lateral scales down to several metres. An example of the latter occurs under conditions of a surface wind which causes the upper layer to circulate in vortices several metres apart (Langmuir circulation, see p. 41). Zooplankton respond to these circulating currents and become aggregated in upwelling and downwelling portions of the vortices, depending on individual swimming speeds and densities. Thus in general on a lateral scale from tens of metres to thousands of kilometres, zooplankton will be dispersed according to the scales of ocean turbulence. Since the latter range from small-scale eddies to very large gyres, the scales of zooplankton patchiness will be correspondingly varied. Where ocean turbulence results in the combined effect of an upwelling component and high primary productivity over an appreciable area, herbivorous zooplankton will generally be present in much larger concentrations than in a downwelling region where there is a low primary productivity. This effect accounts for the large stocks of krill (*Euphausia superba*) in the Antarctic; in contrast, a large area of convergence, such as the Sargasso Sea, is relatively poor with respect to zooplankton production.

The reproductive strategy of zooplankton can result in a patchy distribution, as in the case, for example, of a swarm of copepods which might all release their eggs into the water at approximately the same time. Since all the eggs will behave in approximately the same way, the aggregation of adults will be preserved through to the next generation.

The effect of feeding and predation in forming patchy distributions can be most simply illustrated by the diagram of a predator/prey relationship shown in Fig. 3.1. If in this illustration we consider first that a zooplankton population is the predator, then its abundance over time will vary with the amount of prey (phytoplankton) available; consequently the abundance of the organism will tend to be cyclical since as it feeds, the population increases, which will then depress the phytoplankton—a lower phytoplankton density eventually causing a decrease in the zooplankton. In the second instance, one can assume that the zooplankton is the prey for some higher organism (i.e. fish), then a similar cyclical distribution in zooplankton is generated. The time axes in Fig. 3.1 can

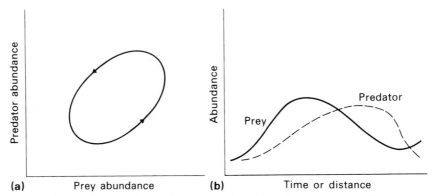

Fig. 3.1. An illustration of variations in the abundance of predator and prey with time or distance. [Modified from Mally E.J. 1969, *Ecology* **50**, 59–73]

be replaced by distance, which would result in spatial instead of temporal changes in abundance (e.g. by assuming that an upwelling occurs at the 0 intercept and that water is being carried away from this point with phytoplankton and zooplankton developing in sequence at some distance).

Since in nature, all the above effects, including both physicochemical and interactive biological effects can operate at the same time, it is not too surprising to find that zooplankton are distributed unevenly in water masses. A few examples of these uneven distributions on different time/space scales in the oceans are given in Fig. 3.2.

The question of what ecological advantages are to be gained from spatial and temporal variations in zooplankton abundance is a subject of considerable

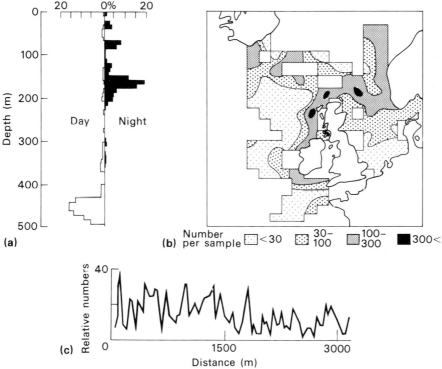

Fig. 3.2. Examples of zooplankton distributions in the marine environment. (a) Vertical position of the copepod *Pleuromamma robusta* during the day and at night, showing strong diel migration; data from the North Atlantic. [From Longhurst A.R. 1976, *The Ecology of the Seas*, pp. 116–140]. (b) Large-scale spatial distribution of the copepod *Calanus finmarchicus* in the North Atlantic. [From Colebrook *et al.* 1961, *Bull. Mar. Ecol.* **5**, 67–80]. (c) Small-scale spatial distribution of the chaetognath *Sagitta enflata*. [From Wiebe P.H. 1970, *Limnol. Oceanogr.* **15**, 205–217]

interest in a number of theories on predator/prey interactions and strategies for species survival. The particular case of vertical migration is considered in Chapter 9. In the case of lateral patchiness, the accumulation of prey organisms gives rise to a higher food intake per unit effort; however, the low levels of prey items in between patches of plankton leads to survival of the population since predators will avoid areas sparse in plankton when dense accumulations are more easily grazed. Thus the occurrence of patchy plankton distributions is an integral part of the ecology of aquatic ecosystems including the way in which the whole pelagic ecosystem maintains itself without exhausting any one part of the food web.

3.1.3 DIVERSITY AND AFFINITY IN THE ZOOPLANKTON COMMUNITY

The diversity of a community of organisms is an expression of the number of different species present in relation to the total number of organisms. Thus a simple expression of diversity (d), called 'Margalef's Index' is

$$d = \frac{S-1}{\ln N} \qquad\qquad (Equation\ 3.i)$$

where S is the number of species and N is the number of organisms. Another expression often employed is the Shannon-Weiner function, H, such that

$$H = -\sum \frac{n_i}{N} \log_2 \frac{n_i}{N} \qquad\qquad (Equation\ 3.ii)$$

where n_i is the number of the 'i'th species and N is the total number of species.

In general the diversity of zooplankton populations increases under conditions of physical stability of a water mass. Thus tropical zooplankton communities are more diverse than temperate communities because the large change in the physical environment from summer to winter in temperate latitudes allows for the survival of fewer species. Diversity also generally increases with the age of a community, more species having become established over a period of time, providing environmental conditions remain constant.

Affinity among zooplankton represents the degree to which certain species of zooplankton are invariably associated with one another. Thus if the presence of one zooplankter is generally accompanied by the presence of another, it may be said that a high affinity exists between the two species. Using a device known as a Hardy plankton recorder which collects continuous samples of zooplankton from commercial vessels, it has been possible to show that certain groups of zooplankton are characteristic of certain water types. For example, copepods from the North Atlantic can be described as oceanic or neritic; among the oceanic a further division can be made into northern temperate species associations and southern subtropical species associations. This idea that certain species characterize water masses has been taken a step further to the concept of 'indicator species' which are supposed to be representative of particular water properties. However, outside of the rather broad division of zooplankton as described above, it becomes less clear what the presence or absence of a particular 'indicator species' specifically indicates. Thus the term has tended to have limited value in a practical sense.

Physical/chemical effects, of the type indicated by data from the Hardy recorder surveys described above, are not the only factors establishing particular associations of zooplankton species. Biological interactions may also lead to the establishment of certain zooplankton groups. This has been clearly demonstrated in lakes where predation by fish in a New England lake was shown to cause a change in zooplankton association from a community dominated by large *Epischura* and *Daphnia* before fish were introduced, to one dominated by small *Bosmina* and *Tropocyclops* after fish were introduced (Brooks & Dodstatter 1965).

52

An expression for the affinity of zooplankton has been given as:

$$I = [J/(N_A N_B)^{\frac{1}{2}}] - \tfrac{1}{2}(N_B)^{\frac{1}{2}} \qquad\qquad (Equation\ 3.iii)$$

where N_A and N_B are the total number of occurrences of species A and B respectively (chosen such that $N_A \leqslant N_B$), and J is the number of joint occurrences. Using an arbitrary level of significance (e.g. $I \geqslant 0.5$), affinity between groups can be expressed as the proportion of paired species having a level of affinity equal to or greater than 0.5. Thus if there are ten species in one group of plankton and two species in another, the occurrence of three species pairs between groups with an affinity index $\geqslant 0.5$ would indicate a relatively low (3/20) species pair connection between these two groups. Using this technique, Fager and McGowan (1963) showed that certain groups of zooplankton could be identified in the ocean. However, the associations were not a simple function of the physical/chemical properties of the water, but were probably (as in the earlier case of 'indicator species') attributable to abiotic and biotic factors yet to be discovered.

3.1.4 THE GLOBAL DISTRIBUTION AND MAGNITUDE OF SECONDARY PRODUCTION

The secondary production of the hydrosphere can be directly correlated with the primary production. However, the relationship is curvilinear since at higher primary production levels there is a decrease in the ecological efficiency in transferring primary production to zooplankton (see Fig. 3.5). Cushing (1971a) showed that there was at least a three-fold decrease in the ecological efficiency for a ten-fold increase in primary production. His results are summarized in Table 3.2 which shows an approximate global range for secondary production of 15 to 50 mg $C/m^2/day$. Thus, over large areas of ocean, including the deep open waters of the North and South Atlantic and Pacific, the annual secondary production would amount to 5 g C/m^2 or less; by contrast in certain upwelling areas such as the Peru Current the annual production of zooplankton may reach about 20 g C/m^2.

Table 3.2. The approximate range of marine secondary production. [Based on Cushing, 1971a]

	Approximate primary production (g C/m²/day)	Transfer efficiency (%)	Approximate Secondary production (mg C/m²/day)
Oligotrophic (e.g. open ocean)	0.1	15	15
Mesotrophic (e.g. continental shelf)	0.3	10	30
Eutrophic (e.g. coastal or upwelling)	1.0	5	50

The daily figures of secondary production given in Table 3.2 are an average for the year and do not represent the maximum production per day. Thus during a zooplankter's life cycle there is generally a period of maximum growth in which the daily production per m^2 may amount to several hundred milligrams of carbon. However, this is followed by periods in which the organism may not

53

grow at all, or even decrease its biomass by living off its storage products during a period of relative hibernation. In addition, the relationship between primary production and secondary production (indicated in Table 3.2) is in fact difficult to establish as discussed in section 3.2. In some cases, for example, a very high primary production in coastal areas may not be grazed at all by pelagic zooplankton but sediment out where it would serve as food for benthic producers.

3.2   Dynamic properties of the zooplankton

3.2.1   FEEDING

Zooplankton feeding mechanisms can be divided into two general types which may be called either filter feeding or raptorial feeding. The former implies that the animal possesses some kind of 'basket' or 'screen' with which to remove small particles (see pp. 167–9); the latter implies that the animal has structures which enable it to seize prey items. The two methods of feeding are not mutually exclusive and among some secondary producers there is the ability both to filter-feed on small particles and to seize larger particles, depending on the availability of different sized prey items. An illustration of various examples of zooplankton with different feeding abilities is shown in Fig. 3.3.

Feeding has been traditionally measured as the volume of water filtered (or volume 'swept' clear) by a planktonic animal. Under experimental conditions, since the number of prey particles remaining (C_t) after a time period, t, is a function of the number originally present (C_0) and the filtering rate (F), one can write an expression for F such that:

$$F = \frac{V(\log C_0 - \log C_t)2.303}{t}$$

(Equation 3.iv)

where V is the volume of the fluid in which the animal is contained. However, two problems with such an expression are that F is assumed constant throughout the time period t, and secondly, in cases where an animal may feed raptorially, there is no measure of the animal's ability to select a particular prey item. In the former case, animals may alter their filtering rate as they become satiated or as the prey density declines; in the latter case, animals may select large prey items, thus consuming more food than indicated by a change in number of prey items alone. However, in general the filtering rate of an organism will be determined by its body size, and the concentration of food particles. Thus for a small copepod nauplius the amount of water filtered per day amounts to a few millilitres while for a large adult copepod the filtration rate may be closer to a litre per day. An increase in food concentration depresses the filtering rate, but the actual amount of food consumed generally increases with prey concentration up to a point where the animal becomes physically incapable of consuming more food. Over short time intervals zooplankton also have some ability to adapt the feeding process by increasing their digestive capacity.

Another measure of feeding is the actual ration (R) consumed by an animal and also the degree to which the ration is assimilated (A). This leads to a general

54

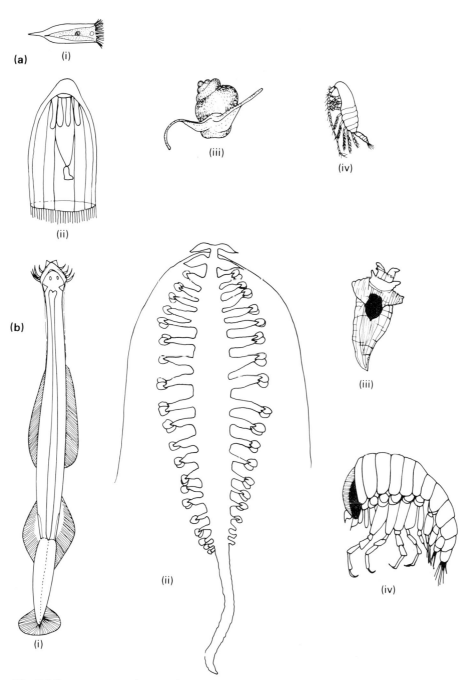

Fig. 3.3 Some representative zooplankton from the marine environment. (a) Filter feeders: (i) Tintinnid (*Parundella* × 100); (ii) Medusa (*Aglantha* × 1); (iii) Pteropod (*Limacina* × 10); (iv) Copepod (*Calanus* × 10). (b) Raptorial feeders: (i) Chaetognath (*Sagitta* × 5); (ii) Polychaete (*Tomopteris* × 5); (iii) Pteropod (*Clione* × 5); (iv) Amphipod (*Hyperia* × 5).

relationship between the growth (G) of an animal, its body respiration (T) and the food intake which can be given as

$$G = AR - T \qquad \text{(Equation 3.v)}$$

3.2.2 GROWTH

Taking the first term (G) in the equation above (3.v), it is obviously necessary to measure growth over some period of time. Thus the growth rate of an animal can be expressed as a percentage increase in body weight per day, in terms of doubling time (t_d) or as the growth coefficient (k). If the animal is assumed to grow exponentially during some portion of its life cycle, then:

$$k = \frac{\log_e(W + \Delta W_t) - \log_e(W)}{t} \qquad \text{(Equation 3.vi)}$$

where ΔW_t is the animal's increase in weight during the time period, t. The growth coefficient (k) is then related to t_d as:

$$t_d = \frac{0.69}{k} \qquad \text{(Equation 3.vii)}$$

where t_d is the time taken for the animal to double its weight. However, zooplankton will only grow exponentially providing adequate food is available during maturation from young to adult stages. During the rest of the year the animal may be inactive and living off body reserves, or in some other part of its life cycle (such as mating or egg laying) when food will be channelled towards these exercises rather than growth. For some of the zooplankton, such as Protozoa and salps, growth has been observed to double the body weight in a matter of a few hours when food is plentiful. However, for organisms having a larger size at adulthood (e.g. the large coelenterates) growth may proceed more slowly resulting in doubling times of days or weeks. Finally, since the process of growth is temperature-dependent, it is apparent that the slowest growth rates of planktonic organisms will be found in colder water, as found near the poles, in deep water and seasonally during the winter in temperate climates. In this respect some arctic zooplankton may require a year or more to reach maturity while similar species in tropical waters can have several generations per year.

Under conditions in which it is possible to standardize temperature and to allow for maximum food consumption, growth of organisms measured in terms of their doubling time (t_d) shows an overall relationship to size. This relationship can be given very approximately as

$$\log t_d = 0.65 \log D + 0.3 \qquad \text{(Equation 3.viii)}$$

where D is the diameter of a sphere in micrometres having a volume equivalent to the volume of the animal and t_d is measured in hours. Using this equation (3.viii) for a temperate euphausiid ($D = 10^4 \, \mu$m) and a small copepod ($D = 10^3 \, \mu$m), one obtains doubling times of about 33 and 7 days, respectively.

3.2.3 ASSIMILATION

Assimilation of food by zooplankton (A in Eq. 3.v) reflects how much of the food captured by an animal is actually absorbed by the body, as opposed to being discarded in the faeces. Assimilation can be expressed as a percentage of the food intake and is generally high (>90%) for carnivorous animals because the biochemical composition of the prey is generally similar to the composition of the predator. Lower values (70–80%) are encountered among herbivores, while the lowest values (<40%) are encountered among detritus feeders where a large amount of the ration is undigestible (see p. 102). As similation has also been found to decline with the quantity of food eaten both among carnivores and herbivores. Thus if food intake is increased by a factor of five, assimilation decreases from over 70% to less than 20%. This gives rise to 'superfluous' feeding in which an animal consumes more food than it can digest. Ecologically this is an important process since the faecal pellets of superfluous feeders will greatly enrich the food available to animals living at greater depths in the water column, including the benthic community.

3.2.4 METABOLISM

The amount of food used for metabolism by an animal is a function of the amount used for such internal body functions as blood circulation, muscle movement, digestion and many other body processes. Metabolism is therefore dependent on an animal's activity but for comparison with other animals, resting metabolism at a specific temperature is generally found to be a function of body weight such that:

$$T = \alpha W^\delta$$

or

$$\log T = \log \alpha + \delta \log W \qquad (Equation\ 3.ix)$$

For many terrestrial animals, δ is considered constant and α depends on the units employed and the experimental temperature used in comparing any group of animals. However, for marine zooplankton living in very different temperature regimes (i.e. boreal, temperate or tropical waters), an adaptation of metabolic processes, indicated by differences in δ, has been found. Ikeda (1970) plotted regression equations of respiration per unit weight against body weight and showed that in different temperature regimes the constants were as follows:

Boreal species	$R = 0.023 - 0.169 W$
Temperate species	$R = 0.357 - 0.309 W$
Tropical species	$R = 0.874 - 0.464 W$

where R is the logarithm of the oxygen consumption in μl/mg body wt./hr and W is the logarithm of the dry weight of the animal expressed in mg.

3.2.5 FOOD INTAKE

From the foregoing discussion it is apparent that because of differences in metabolism of different-sized animals, assimilation of different kinds of food,

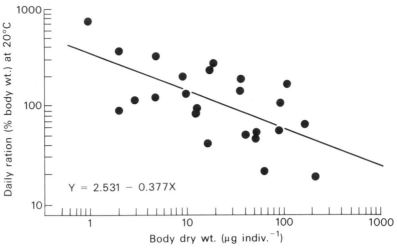

Fig. 3.4. A relationship between daily ration and body weight in various marine planktonic copepods. [Data accumulated by Ikeda T. 1977, *J. exp. mar. Biol. Ecol.* **29**, 263–277].

different growth rates and the effect of temperature on all these processes, the amount of food actually eaten by a zooplankter can be highly variable. However, in general, smaller animals metabolize faster and grow faster than larger zooplankton. Thus their food intake per day is higher as a percentage of their body weight, than for larger zooplankton. This is illustrated in Fig. 3.4 which shows that the food intake per day as a percentage of the animal's body weight for different sized marine crustaceans can vary from well over 100% per day for small animals, down to less than 20% per day for large crustaceans. The partitioning of the food intake between growth and metabolism is illustrated for a nauplius and copepodite of *Acartia clausii* in Table 3.3.

Table 3.3. Distribution of food eaten as a percentage of the body weight in *Acartia clausii*. [From Petipa, 1966]

	Dry wt. (μg)	Daily growth (%)	Daily metabolism (%)	Daily ration (%)
Nauplii	0.09	20.3	98.3	148
Stage V	2.56	8.3	49.8	73

3.2.6 NUTRIENT REGENERATION BY ZOOPLANKTON

In the process of feeding and excreting, zooplankton are responsible for an appreciable recycling of nutrients, particularly the major phytoplankton nutrients, nitrogen and phosphorus. The amount of nitrogen excreted per day by a zooplankter is probably in the range of 2 to 10% of the animal's body nitrogen. This value is difficult to determine, however, because metabolic nitrogen (in the form of ammonia) is excreted along with the undigested faecal material. Phosphorus is excreted as well as nitrogen in an N:P ratio of *c*.5:1. In the sea at least, since organic phosphorus compounds are readily hydrolysed at an alkaline pH, much of the excreted phosphorus appears as inorganic phosphate.

The excretion of metabolic products depends to some extent on the amount of food being eaten, and at high food concentrations the excretion rate of nitrogen and phosphorus may be five to ten times greater than if the animals are in a resting metabolic state.

The turnover of nutrients by herbivorous zooplankton has its counterpart in carnivorous zooplankton, particularly the marine ctenophores and medusae. Because of the voracious appetites and rapid growth rates of these carnivores, large stocks of copepods can be destroyed and nutrients liberated to continue the food cycle of the sea.

3.2.7   THE ROLE OF DETRITUS IN THE FOOD OF ZOOPLANKTON

In Fig. 3.1 it appears that a predator is wholly dependent on living prey for its food. This is generally true but in the case of the herbivorous filter feeders there is reason to believe that at some times of the year their principal source of food may be particulate organic detritus. The chief evidence for this is largely ecological. Thus in studies on zooplankton communities in which researchers have attempted to define a budget for the food requirements of a filter-feeding zooplankton community, it has been necessary to assume that part of the zooplankton diet is obtained from detrital material in the water column, particularly during the winter months when there is very little primary productivity. In coral reef communities additional evidence has been obtained to show that throughout the year, organic material released by the coral may form particles which are utilized by pelagic zooplankton as food. This would help to explain the very tight conservation of energy which must be practised within such communities in order for them to survive under otherwise oligotrophic conditions.

Experimental evidence that filter-feeding zooplankton can exist and grow on detrital material does not generally support the ecological assumptions made above. However, this in part may be due to the way in which the experiments have been conducted. It is difficult to reproduce experimentally the fine particulate detritus of the sea and feed it to zooplankton. In experiments in which detritus was collected from the sea in the absence of phytoplankton, the resulting particulate material was not shown to be an adequate source of food for zooplankton *per se*. However, there are two important exceptions to this experimental observation. One is that the faecal material of copepods has been shown to serve as an adequate food source for other filter-feeding plankton and secondly, if soluble organic material is converted into bacterial biomass, this too is an adequate food for zooplankton. Thus the role of detritus as a food source for zooplankton may be quite heavily linked to either the bacterial cycle of the sea or to the presence of faecal material from other grazers. These sources of organic carbon are included in the general assessment of detritus since they represent organic carbon, other than that of the photosynthetic primary producers. However, also included with the detrital organic carbon are appreciable quantities of refractory materials and it is probably these materials which are not in fact readily digested by the zooplankton, even if they are consumed.

3.2.8 FOOD PREFERENCE IN RELATION TO PARTICLE SIZE OF PRIMARY PRODUCTS

There is a general consensus of results which shows that the very largest herbivorous zooplankton (e.g. the marine euphausiids) graze on large diatom chains and if these are not available, feeding strategy may change to the capturing of small animal prey. Conversely some of the smallest crustaceans (e.g. copepod nauplii) are smaller than the largest diatom chains and among these animals the appropriate prey item is a small flagellate; the latter may be several orders of magnitude smaller than the largest primary producers (i.e. the size range of photosynthetic phytoplankton in the sea, on a microscopic scale from 5 μm to 10 mm, changes over the same orders of magnitude as one observes in the terrestrial environment between primary producers such as grass, up to the largest trees).

Within this general size relationship between herbivorous crustaceans and their prey, there are a number of exceptions which tend to modify the relationship. The first of these is that for less extreme size ranges, there is a within-size relationship which tends to show that the abundance of a prey item will be more important than its relative size, for a size range which differs by no more than a few hundred per cent. Thus copepods of significantly different size have been shown to predate the same prey if it was the most abundant phytoplankton present. Outside of the crustacean community there are numerous examples of relatively large herbivorous zooplankton filtering small prey items, using a finer filtering apparatus than is available to the crustaceans. These include the salps, larvaceans and thecate pteropods, all of which are relatively large organisms capable of filtering out the very smallest phytoplankton. Another exception to a general prey size relationship among planktonic zooplankton are certain crustaceans (e.g. the copepod, *Oithona*) which as a carnivore is capable of capturing other copepods of approximately the same size as itself.

3.3 Production

3.3.1 CALCULATION OF PRODUCTION

Production is defined as the total elaboration of new body substance in a stock during a unit time, regardless of whether or not it survives to the end of that time. Thus in the absence of predation or natural mortality, production is measured by the growth of a population from the egg to the adult stage; since the change in biomass (B) of a large copepod from egg to adult can be as much as a thousand-fold, the potential production (P) to biomass ratio (P/B) could be as high as a hundred (assuming a geometric mean biomass over time of 10). However, under natural conditions there is a mortality in the number of individuals (N) from the beginning of the period of observation to a lesser number (N_t) after time, t. A simple expression of production (P_t) can then be made, assuming a short time interval for t, to give

$$P_t = (N - N_t) \cdot \frac{W + W_t}{2} + (B_t - B)^* \qquad (Equation\ 3.x)$$

where W and W_t are the average individual weights of organisms at the beginning and after time t, respectively, and N and B are the corresponding observed number of individuals and total biomass of the population. In this expression, $B_t - B$ represents the production surviving after time t and the rest of the expression represents the production lost due to unspecified mortality during the same period.

If P_t is determined at different points in the life cycle of a zooplankter it will generally be positive during periods of maximum growth and negative during periods of hibernation (e.g. for some species at depth during the winter). Thus the P/B ratio will depend on the time of observation; for comparison with other populations it is best to consider P/B ratios on an annual basis. This can be done by summing the separate estimates of P_t to give:

$$P_{\text{Total}} = P_1 + P_2 + P_3 \ldots P_i \qquad\qquad (Equation\ 3.xi)$$

and expressing P_{Total} as a ratio of the average standing stock $(\bar{B})$ throughout the year. When this is done the P/B ratio for many zooplankton populations lies in the range of 10 to 40 (in contrast to approximate P/B ratios for phytoplankton of 300 and for fish of 1, or less).

Table 3.4. The production of *Acartia clausii* as determined from different size groups of individuals. [From Mann, 1969]

Size groups $(mg \times 10^{-3})$	Number/m³	Individual daily growth $(mg \times 10^{-3})$	Daily production $(mg \times 10^{-3})$
0–2.5	850	0.09	76.5
2.5–10	250	0.45	108.0
10–20	87	1.20	104.0
20–30	35	1.68	58.8
30–50	47	1.20	56.4
50–70	40	0.47	18.8
		Total	422.5

The method of calculating production as described above can be readily applied to a zooplankton population which has only one generation per year (i.e. a single cohort). It becomes more difficult among ephemeral species where several cohorts may be present in the water at any one time. The problem of determining the growth rate (using either Eq. 3.x, or exponential kinetics) must

---

*This expression (3.x) assumes a linear function for the change in the biomass and number of organisms over a short time interval, t. A better approximation of growth and mortality may be obtained by assuming an exponential function such that production, P, is

$$P = \left(1 + \frac{m}{g}\right) \Delta B$$

where m is the mortality coefficient of organisms
$$N_t = Ne^{-mt}$$
and g is the growth coefficient of biomass
$$B_t = Be^{gt}$$
and ΔB is $B_t - B$. For a case in which $\Delta B = 0$ and hence $g = 0$, the expression for production becomes

$$P = mBt$$

where t is the time interval of the observation as previously. It should be noted that in both these estimates of production it is assumed that no reproduction (only growth of individuals) occurs during the time interval, t.

then be resolved either by measuring the growth rate of individual zooplankton experimentally or by separating out different sized organisms from the population and determining the number and weight of each size class. An example of the latter calculation for the daily production of different sized specimens of the marine copepod *Acartia clausii*, is given in Table 3.4. Mann (1969) gives details of other methods of measuring and calculating production.

3.3.2 THE RATIO OF PRIMARY TO SECONDARY PRODUCTION IN DIFFERENT AQUATIC ECOSYSTEMS

Since the production of zooplankton is inherently related to the growth of individuals, and since growth is related to the quantity of food eaten (section 3.2.5) it would seem that production should increase with the availability of food, i.e. from oligotrophic to eutrophic conditions. In a rather limited sense this appears to be true, but there are a number of depensatory mechanisms which may make it difficult to observe this apparently simple connection. For example, as primary production increases, there may be a resultant increase in the biomass (B) of zooplankton (if they are not predated). However, this larger biomass of zooplankton may not necessarily produce more zooplankton tissue per unit time (dB/dt) than was being produced by a smaller initial population (the excess food being used in the former case to meet the larger metabolic requirements of the population and not for growth). In another example, an increase in ration (e.g. phytoplankton for a herbivorous zooplankton population) may be of the wrong species composition for an endemic population of zooplankton and hence the zooplankton may utilize the phytoplankton less

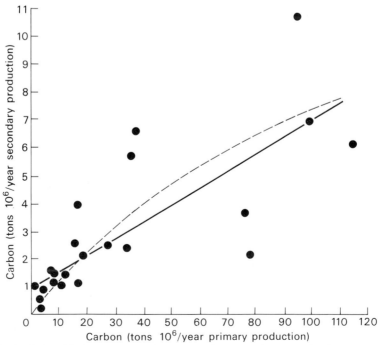

Fig. 3.5. A relationship between secondary production and primary production. [Data accumulated by Cushing D.H. 1971b, *Adv. mar. Biol.* **9**, 255–334]

efficiently than before there was an increase in ration. Thus, even under experimental conditions it has been difficult to establish a direct relationship between production and available food supply. However, on an overall basis, where lakes and sea areas have been allowed to adjust to sustained conditions of moderate enrichment (either from the land into lakes or in upwelling areas of the oceans) an increase in the primary production of food organisms can be observed to support a higher production of secondary producers (zooplankton). Such a positive relationship is shown in Fig. 3.5 which represents a summary of data on primary and secondary production measured in a number of different sea areas of the world.

In Fig. 3.5 it will be seen that while the relationship is positive, the best fitting line is convex and indicates that although primary production causes an increase in secondary production, there is also a general loss of efficiency in transferring from phytoplankton to zooplankton as primary production increases. This loss of efficiency has been investigated under experimental conditions in which several ponds were treated with different levels of nutrients and the resulting levels of primary and secondary production were compared over a period of time. Table 3.5 shows the results of this experiment in which the ratio of herbivorous zooplankton production to primary production is clearly shown to decrease with enrichment from low to high nutrients (see p. 53).

Table 3.5. The ratio of herbivore to primary production in experimental ponds. [Taken from Pederson *et al.* 1976].

Experimental pond	Herbivore production / Primary production
High nutrients	0.07 to 0.05
Medium nutrients	0.41 to 0.06
Low nutrients	0.56 to 0.20

3.3.3 THE RELATIONSHIP BETWEEN SECONDARY PRODUCTION AND THE PRODUCTION OF FISH

The problem of establishing a direct causal relationship between zooplankton and fish production has been equally difficult to that with zooplankton and primary production. Not only does one have similar problems as described in the previous section, but there are added complications when dealing with data on fish. In the case of fish their movement over wide areas, together with the fact that statistics on fish are generally collected for purposes of commercial catch (Chapter 8), and not in relation to food supply, makes it very difficult to discover the form of the relationship between secondary and tertiary producers. Added to this problem, although the zooplankton form the basis for food for most fish stocks, the complicated nature of the food web leading to some fish stocks introduces too many unknowns between secondary production and the final commercial harvest of fish. It can be shown in general that in oceanic areas where there is a high primary production there is also a high fish production, and in most cases it must be assumed that this is mediated by having a high secondary production. However, the best illustration of the effect of secondary producers on fish production can be found by an inter- and intrasystem com-

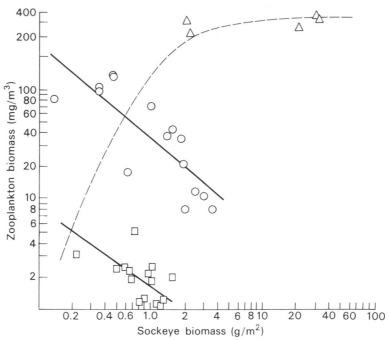

Fig. 3.6. Relationship between the biomass of zooplankton and sockeye salmon within systems (solid lines) and between systems (dashed lines). △ Dalnee. ○ Babine-Nilkitkwa. □ Owikeno. [Data accumulated by Brocksen *et al.* 1970, *Marine Food Chains* pp. 468–498]

parison of certain lakes where it is known that because the fish are isolated in these waters, they have to be directly dependent on the endemic secondary production. This is illustrated in Fig. 3.6. The figure shows that the zooplankton biomass within Lake Owikeno and Babine Lake decrease with the standing stock of sockeye salmon. Assuming that the production of zooplankton is constant in each lake, then the depressed biomass of zooplankton in the presence of more fish is a direct relationship between secondary production and fish production. Further, it is apparent that the intercepts on the zooplankton and fish axes of the graph show a greater standing stock of both organisms in Babine compared with Lake Owikeno. Thus on an interlake basis, the more eutrophic lake (Babine) supports more secondary production and more fish than the oligotrophic lake (Owikeno). These intra- and intersystem comparisons can be carried further to Lake Dalnee which is the most eutrophic lake of the three. There one sees that the biomass of fish is greater for a greater biomass of zooplankton, which confirms the previous interlake relationship. However, within Lake Dalnee, an increase in biomass of sockeye salmon does not depress the abundance of zooplankton. In this case there must be some other factor controlling salmon production (e.g. possibly space requirements); such an independence of fish production on zooplankton would only be possible at very high levels of secondary production. From the comparison of these three lakes it is possible to approximate the overall nature of the zooplankton/fish relationship as indicated by the dotted line in Fig. 3.6. This shows that there is a good general relationship between increased biomass of zooplankton and increased biomass of fish, up to some limiting value of zooplankton biomass where the fish are apparently getting all the food which they need to survive, but where some other factor is controlling their biomass.

In conclusion, it can be said that a direct relationship exists between fish production and secondary production so long as no other factors of either a density dependent or density independent nature are introduced to affect this dependence. If such factors are introduced (e.g. a density independent effect of temperature causing high fish mortality) then the relationship between secondary production and fish production will not hold. Further, if the fish population is removed in its trophic position, such as from being a planktivore to being a piscivore, then the direct dependence of the latter population on secondary production may be extremely tenuous.

Another aspect of the relationship between secondary producers and fish is the food preference of fish for certain types of zooplankton. Thus, in general the large 'jelly' zooplankton community of the sea (including the true jellyfish, salps and ctenophores) is not extensively eaten by fish. In contrast the crustacean zooplankton community is extensively grazed. However, within the crustacean community there are certain food preferences which are partly related to size and abundance of the particular zooplankton. In the feeding of mackerel, for example, it has been shown that these fish will select relatively large prey items down to a concentration where the large zooplankton are less than 10% of the total zooplankton biomass. At this point the fish is capable of changing its feeding strategy to collect very small zooplankton by filtering them on their gill rakers (O'Connell & Zweifel 1972). The preference of fish towards other zooplankton may also be related to the colour of prey items, presence of spines, speed of escape and other strategies which have evolved among the zooplankton in order to ensure the survival of at least some members of the species.

3.4 Conclusions

The zooplankton of aquatic communities form an ecological group which is difficult to compare with any similar group of animals in terrestrial ecology. The herbivorous filter-feeding zooplankton are in the same trophic position as the large terrestrial herbivores, yet their strategy for survival is completely different, being orientated toward an r-selected survival as opposed to being K-selected (see Prologue for definitions of r and K). Thus the occurrence of zooplankton in both time and space can vary between a superabundance of organisms to a virtual absence of the same species.

In size, zooplankton range over three to four orders of magnitude; thus although the number of species is large, it is possible to view the community as a continuous size spectrum of animals in which the smaller organisms tend towards herbivorous feeding (e.g. many, but not all, copepods) and larger organisms (e.g. fish larvae, coelenterates and ctenophores) tend to be carnivores. By using this concept of size distribution it is possible to relate many of the dynamic processes of the zooplankton community to size functions. Thus expressions for metabolism, feeding, growth, locomotion and the intrinsic rate of population increase can all be represented as a general function of size, providing animals are compared under reasonably similar conditions. The writing of various functions and research on evaluating the coefficients in these functions has led to the development of trophodynamic models of the zooplankton community. However, because of mathematical and logistic restrictions, such models can only be used to examine further the complexity of plankton relationships. This is particularly true when one considers, for example, the great variety of life cycle strategies employed among organisms of the plankton community; no model could be written which could encompass the infinite resourcefulness of nature to provide organisms wherever an opportunity presents itself as a place for life to prosper. Thus in the study of zooplankton communities one tends at times toward a holistic view of community interactions and at times toward the reductionist view in order to try to explain the unexpected occurrence of a particular event. It is quite apparent under such circumstances that the various branches of science used to study zooplankton, from their taxonomy to sophisticated mathematical models, are all useful in explaining how this community functions.

4

Primary Production of Benthic and Fringing Plant Communities

J.M. TEAL

4.1 Introduction

Fringing plants grow around every body of water. They vary from a few lichens and attached diatoms growing on steep rocks on the shores of a deep lake to extensive marshlands in which open water may be almost entirely lacking. In such a continuum the importance of the fringing plants will vary from insignificant to providing virtually all of the primary production within the aquatic system. In this chapter we will be mostly concerned with those aquatic systems in which the fringing plants make an important contribution, i.e. systems where circumstances allow substantial growth of attached plants as distinguished from planktonic plants, at least in the immediate vicinity of the shore.

We must consider many different types of plants but they can be grouped according to their manner of growth and the way in which they respond to their environment. At one extreme are the microphytes, single-celled algae or microscopic filaments of cells living on surfaces of the bottom or on larger plants. Lichens might reasonably be considered together with these as a group of plants conforming very closely to the bottom, projecting very little into the water. Another group consists of what are commonly thought of as seaweeds. These are macrophytic algae, usually projecting up into the water, sometimes for considerable distances. They have structures for holding themselves onto or in the substrate but they typically have little rigidity. They depend on the support of the water to stay erect and intercept light efficiently. Either they are held up and moved about by water movements, or they may have air bladders which float their fronds up into the overlying water.

The third group is composed of vascular macrophytes, both herbaceous and woody. These are all rooted plants. Roots not only anchor them to the substrate but are specialized for uptake of nutrients and water. They are typically strong enough to hold themselves erect, the woody species carrying this to the extreme of building the long lasting structures for which they are named that hold their photosynthetic leaves up into the air and light.

There is a great diversity of plants in these three groups. For example, diatoms living on the surface of the bottom, or attached to higher plants growing on the bottom, are closely related to the planktonic diatoms discussed in Chapter 2. Seaweeds are most characteristic of rocky marine coasts where we can find rock weeds or wracks (*Fucus* and *Ascophyllum*) intertidally and kelps and red algae such as Irish moss (*Chondrus*) growing below the tide zone. In freshwaters the comparable position is typically occupied, in a series of zones from the shore to deeper water, by:

1 Emergent plants such as sedges and cattails which project through the water surface.

2 Plants with floating leaves such as water lilies.

3 Submersed rooted plants such as pond weeds and milfoil but including a few freshwater macroalgae such as the stoneworts (Charales) abundant in marl lakes.

Totally submersed vascular plants are not characteristic of the sea but a group of grass-like plants, the sea grasses occur along soft-bottomed temperate and tropical coasts. Free-floating plants such as water fern and duckweed can occur extensively throughout all these zones. In quiet shallow waters they may cut off enough light to inhibit greatly the growth of the submerged plants. Marshes, usually more like flooded grasslands than open water, are of similar appearance in both fresh and salt water habitats. Swamps are covered with a great variety of plants depending on the length and period of year when they are wet. The cypress swamps of the southeastern United States, the vast swamps of the Rio Negro in northern South America or the Congo in Africa, and the mangrove swamps found along tropical coastlines round the world are all good examples of situations with shallow water where trees have been able to form extensive wetlands.

Bogs are another type of wetland in which the dominant plants are mosses, typically *Sphagnum* species. They are very extensive especially at high latitudes, often without associated open water. Quaking bogs occur as littoral fringes round relatively deep small lakes. *Sphagnum* encroaches on the water surface as a floating or semi-floating layer which is colonized by sedges and acid-loving plants such as leather-leaf, *Chamaedaphne*, and eventually by bog trees including black spruce and larch. Because the bog is not resting on firm ground it quakes when stepped upon, hence its name.

4.2　Levels of production

4.2.1　PRODUCTION MEASUREMENT

Before we consider the productivity of fringing plant communities, we have to say a few words about the ways in which production is measured. Although most measurements of algal production, including the microplants within the fringing communities, are made by carbon or oxygen exchange or estimated from chlorophyll concentration, most measurements with larger plants are done by harvesting and weighing the plants. Gas-exchange measurements have been made on fringing communities and can give the best measures of actual carbon fixation available. But even here there are problems. If the gas-exchange measurement was made by enclosing a part of the whole community, the respiration will include that of the animals and microbes as well as that of the plants and the production will be that of the community as a unit rather than that of the plants only. The biggest problems with harvest methods lies in knowing how much of the plants under consideration were harvested and when. Weights may be reported as wet weight which would be very misleading if trying to compare the watery tissues of submersed freshwater plants with woody stems. More often dry weights are reported but that is subject to error depending on the amount of ash in the plant tissues. While many plants have between 5 and 10% ash in their dry tissue, some such as diatoms, calcareous algae and some vascular plants of saline environments contain from 50% to as high as 90% ash. The most useful comparisons are in terms of organic matter (ash-free dry weight) or of carbon. I will use the former here but the student should remember that the comparisons made below are inexact because of the inaccurate conversions that must be used to put the available data into comparable form.

There is a further problem in making comparisons because of the ways in which the harvests were made. Frequently the underground parts of the plants are not harvested though these may represent more than half of the total plant production. A common method of measuring standing crop is to harvest the crop at the period of its maximum biomass and report that as production. This will be an underestimate because the technique misses leaves or stems that have grown but died and disappeared before and after the measurement was made. A study made of one warm-temperate marsh reported that annual production was $9\frac{1}{2}$ times the maximum standing crop present at any one time in the year. Production measured as maximum standing crop will also be an underestimate if different species reach their maximum biomass at different seasons as often occurs when the vegetation consists of a mixture of annual and perennial species. A freshwater marsh in New Jersey was dominated by *Acorus* and *Peltandra* in June, *Impatiens* in July, and *Zizania*, *Bidens*, and *Polygonum* in August. The maximum above ground standing crop at any one moment, which would be reported as plant production in many studies, was 840 g/m^2. But if the maximum standing crop for each species is summed regardless of when it occurred, then annual production would rise to 1200 g/m^2. Finally when individual leaves were followed so that allowance could be made for their death and disappearance during the growing season (560 g) and root production was included, the estimate became 2100 g/m^2 or $2\frac{1}{2}$ times that reported by the simplest technique.

In plant communities where because of a uniform environment or the longevity of the plants there is little seasonal change in biomass it may be necessary to follow the growth of individual leaves and stems to get useful data. This is illustrated by the technique of punching holes in kelp blades to follow the rate of their renewal (Fig. 4.1).

Besides the already-stated difficulties in knowing what is the real primary production or of making comparisons with most available data there is the further problem that most aquatic plants at some times during their growth secrete or leak organic matter from the plant body into the water. Values ranging from 1 to 10% of the rate of photosynthesis have been reported and these may become much greater (over 40%) during senescence of the plant or of even just an individual leaf.

4.2.2 PRODUCTIVITIES IN NATURE

Table 4.1 lists representative values for the production of fringing communities. Values can be small wherever the system on which the measurement was made

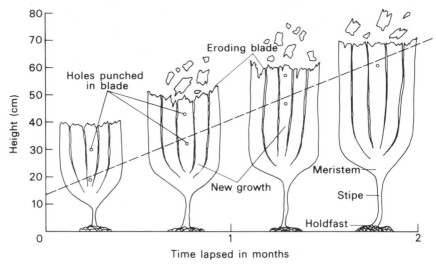

Fig. 4.1. Diagram illustrating the hole-punching method used to monitor the growth of the kelp *Laminaria*. The dashed line shows movement of a punched hole over a 2-month period. Although the hole has moved 40 cm from its initial position 20 cm above the base of the blade, the plant is only 25 cm taller due to erosion at the tips. In winter, plants tend to get shorter, when erosion may be faster than new growth. [After Field *et al.* 1977, *S. Afr. J. Sci.*, **73**, 7–10]

was less than ideal for the plants. The more interesting point is how large the values can be on an areal basis. It would appear that values of about 2000 g organic matter can be produced per year on one square metre by the plants of the fringing communities. In the better conditions the values rise to over 5000 and in the highest reported value to 15 kg. This latter value, reported for papyrus swamps in tropical Africa, was estimated by measuring CO_2 uptake over a short period and extrapolating to a full year. The rate for the short term measurement was not much higher than has been reported for similar emergent plants growing in temperate marshes. The much higher annual production results from the continuous growing season in the tropics for plants that are not limited by a dry season.

We can see how high these rates of production are by comparing them with the highest values from agriculture where the plants are carefully fertilized and weeded to assure maximum growth. Sugar cane can achieve about 8500 g organic matter/m^2/yr, with maize (*Zea mays*) second at about 3500. World averages for these crops are much lower, about 3600 and 900 respectively. Among uncultivated plant systems, rain forests rank highest with production of up to 5000 g/m^2/yr while temperate plant communities are more likely to exhibit values of less than 1000.

Even though the rate of production per unit area may be very high the overall contribution of fringing communities to aquatic production will depend on their area in relation to that of the open water. A rough estimate puts the area of freshwater wetlands at two million square kilometres. The area of salt wetlands is less well known but is probably less than 10% of that of the freshwater. So in this chapter we are considering an area of less than 0.5% of the

Table 4.1. Annual production levels of fringing benthic plant communities. Average levels are compiled from all available data and represent a range of environments though data from very unfavourable sites is probably not included. The maximum values are the highest that could be located. The data are given in g organic matter (ash-free dry wt.) per m^2 per year.

Fringing community type	Average production	Maximum production
Bog	900	1500
Marshes, freshwater		
Typha, cattail	2700	3700
Carex, sedge	1000	1700
Phragmites, reed	2100	3000
Cyperus, papyrus	7500	15000
Freshwater, tidal	1600	2100
Marshes, salt water tidal		
Spartina	2500	6000
Other grasses	1500	8500
Salicornia	3000	
Swamps, freshwater		
Bog, spruce	500	
Cedar, Cypress	4000	
Hardwood	1600	
Swamps, salt water		
Mangrove	3000	
Benthic microphytes		
Freshwater, lake	170	1560
Freshwater, spring	3000	3500
Marine	200	850
Submerged macrophytes		
Freshwater, temperate	650	1300
tropical		1700
Marine		
sea grasses, tropical	2000	8500
macroalgae, kelp	3400	4250
rockweeds	750	2100
green algae, tropical		4000
Floating, freshwater		
Duckweed	150	1500
Water hyacinth (*Eichornia*)		3300

total land area on earth or 0.6% of the area covered by water in both oceans and lakes. But because of its high rate of productivity this minute fraction of the earth's area contributes on the order of 2.5% to total productivity. It is even more significant than these figures alone suggest because it is so highly concentrated in shallow water. This makes it valuable as a source of food for aquatic animals, as a refuge area especially where macrophytes are abundant, as a plant nutrient source and for other reasons which will be discussed in other chapters.

Fringing versus planktonic production

The relative significance of fringing production will depend on the area of the water body available for growth of these plants, as well as on their productivity but not necessarily that of the water body as a whole. Most of Lake Tahoe (Table 4.2) is too deep for fringing plants to grow, it has little soft bottom on which emergent plants could grow, and the waters are poor in nutrients. The result is that only 1% of the production is due to fringing plants which in Lake Tahoe are microphytic algae growing on rocks. At the other extreme we

Table 4.2. The percentage contribution of fringing to total production in lakes with a range of areas, depths and total production.

Lake	Fringing production / Total production (%)	Area	Mean depth (m)	Mean total production (g/m²/yr)
Tahoe	1	499 km²	313	107
Opatovicky Pond	16	24 ha	2	910
Lawrence	75	5 ha	6	360
Marion	98	31 ha	2	106

find lakes like Marion, shallow with a soft sediment in which macrophytes can cover the bottom. Opatovicky Pond is fertilized and has a much higher overall productivity but less is contributed by the fringing plants which cover only a fraction of the pond's total area. Lawrence Lake is a small, shallow marl lake. The contribution of fringing communities to total production is clearly not related to total productivity nor is it associated with area except that very large lakes or seas have relatively little shoreline.

Relative importance of different types of fringing plants

The obvious plants in most fringing communities are the macrophytic algae or the higher plants. We naturally think of them when considering the importance of these communities but substantial parts of the productivity may be due to the microphytes growing either directly upon the substratum or upon the large surface area often provided by the larger plants (Table 4.3). By growing upon the macrophytes, microscopic plants also avoid some of the shading due to their large relatives. The microphytes may even benefit from secretions from the leaves of their hosts, but the relationship is not parasitic and is mostly one of simply finding suitable surface upon which to grow. In some cases, e.g. Marion Lake, the algae seem to do better living on the mud or sand under the plants than on their leaves and stems in spite of the fact that there is much more surface available in the latter habitat. Probably in cases like in a tidal marsh the light or drying conditions up on the larger plants are less favourable than those found on the substratum under them. For whatever reason, the microscopic plants can vary from relatively insignificant to considerably more important than the large plants.

Table 4.3. Relative importance of microphytes and macrophytes in a variety of marine and freshwater fringing benthic communities.

Type of fringing community		Microphyte productivity (g/m²/yr)	Microphyte / Macrophyte production (%)
Lawrence Lake	mud algae	45	1
New Jersey salt marsh	mud algae under grasses	26	5
Opatovicky Pond	epiphytes	77	13
Thalassia bed, Florida (sea grass)	epiphytes	425	20
Lawrence Lake	epiphytes	80	22
Sea grass bed	mud algae under grasses	20	72
Marion Lake	on sand under macrophytes	66	224

4.3 Controls on productivity

Fringing communities can be among the most productive natural systems on earth and often exceed the productivity of managed agricultural systems. The reasons for this high production lie in the good supplies of both water and nutrients available at these sites. A vascular marsh plant is growing in a soil with continuous supply of liquid water. The soil is often continuously enriched by erosion from the watershed, and high in nutrient supply. It is also usually anoxic making it a suitable environment for nitrogen fixation. The tops of the plants are in the air which supplies them with carbon dioxide for photosynthesis and oxygen for nighttime respiration. Diffusion of these gases in air is 10 000 times more rapid than diffusion in water and air is much more fluid so that even though the concentration of carbon dioxide is only 0.03% in air the supply is usually sufficient to supply all the plants needs. Oxygen is 20% by volume of air, 210 ml/litre, while it is typically less than 10 ml/litre in even well oxygenated water, so the requirements for respiration may be even more important than the CO_2 supply.

Tidal marshes have the further advantage that not only is the water moving, continuously renewing the nutrient supply and removing wastes from the system, but with the rise and fall of water level there is a continuous flushing of the soil bringing the benefits directly to the plant roots.

Non-vascular plants are in a different situation since although they have holdfasts to attach themselves to the substrate they lack roots which can absorb nutrients from pore waters even when they grow on soft sediments. They absorb nutrients through their entire surface and depend upon water movements to continuously renew their supply. Water movements must also supply them with sufficient oxygen. Since they can use the water to support them they can build much less non-productive support tissue and devote more of their efforts to photosynthetic structures. The result is that in the case of the large seaweeds, the productivity of their stands is comparable to that of vascular plants. In cases like that found in springs with a large flow of nutrient-rich water, microalgae also achieve very high levels of production.

4.3.1 SEASONALITY

Temperature and light limit production especially in areas where these factors vary seasonally. Freshwater and intertidal plants can be frozen into ice and completely dormant especially in many northern areas. But the effects of cold are not as obvious as might be expected. Plants actually encased in ice are dormant. Plants subject to ice damage such as the tidal marsh grasses in northern marshes where massive blocks of ice shear off all above-ground growth have all their perennial parts down in the frozen mud. But plants living deep enough to be completely under the ice may actively grow during the coldest parts of the year. Kelps growing subtidally along the coast of Eastern Canada exhibit substantial production of new fronds during midwinter when water temperature is close to 0°C. Light is minimal at that season but other conditions are advantageous. Nutrients have been regenerated in the water and their concentration is at a maximum. The phytoplankton have not yet begun to bloom so they are neither competing for nutrients nor shading the kelp. By being adapted to grow efficiently at low light levels, the kelp are able to make use of the available resources to achieve production as high as that of other fringing communities that are active during the long days of summer.

At times the temperatures in some fringing communities may get so high that growth is inhibited. This is most likely to occur in intertidal algae during exposure to intense midday sun or in sheets of floating plants such as duckweed which can intercept all the sunlight in a very thin layer. In both cases the ability of the plants to cool themselves may not be sufficient to prevent heat damage. Probably the restricted air movement just above the plants on still days in intense sun does not allow rapid enough water evaporation to provide the necessary cooling. A grass leaf standing up into the air is much more efficient at cooling under these conditions and the temperature in a marsh grass might not rise above that of the surrounding air while that of the duckweed floating nearby would rise enough for damage to occur.

Change in light through the seasons obviously limits production in winter but can also inhibit some plants by being too intense in summer. In a multi-layered stand of macrophytes though production may be limited at high light in the topmost leaves or fronds, the lower layers will increase production at higher light levels. The overall productivity of the stand increases steadily with increasing light. However, microphytes on the surface of the mud may be inhibited by intense summer light. Salt marsh algae living on mud flats in Georgia migrate to the mud surface at low tide but are under a thin layer of mud at high tide due both to their own movements and deposition of water-borne sediments. The result is that during the summer at low tide the light is too intense for optimum photosynthesis and most production occurs at high tide when the plants are shaded by both a thin layer of mud and the overlying water. During winter with less intense sunlight maximum production occurs at low tide when the algae are lying on the surface exposed to full sunlight. The algal migration does not keep the plants at an optimal light level so it must have developed in response to disturbance by the rising tide, higher nutrient concentrations in deeper sediment layers, or for other reasons, but the result is a more or less constant level of production throughout the year.

4.3.2   WATER SUPPLY

Many fringing benthic communities are well supplied with water through the year, which is one of the reasons for their high production. But in other cases water is definitely limiting at times. The most obvious case of varying water supply is intertidal. Algae living high in the intertidal are subject to the most prolonged drying, but generally these short periods of desiccation seem to stimulate growth rather than inhibit it. Rockweeds increase both their photosynthesis and their elongation by as much as 80% in the first couple of hours after exposure to air. With longer desiccation both processes slow down. The later slowing is probably related to water loss but the initial acceleration is more likely related to temperature increase upon exposure to air.

In swamps and marshes which are inundated unequally throughout the year lack of water can also limit productivity during dry periods. Water loss by transpiration from the great surface area of leaves projecting through the water surface in a marsh is several times greater than evaporation from the surface of an equivalent area of water. This drying potential can speed up the arrival of conditions sufficiently droughty to limit production.

Finally water availability can limit production in salt marshes and swamps. Vascular plants lose freshwater by transpiration, i.e. water evaporates from the stomata on their aerial parts. Any salts that were dissolved in that water remain behind. The stomata must be open to permit uptake of CO_2 from the air and typically several hundred molecules of water are lost for every molecule of CO_2 absorbed. But salt marsh plants must obtain water from the sea water solution surrounding their roots. They have adaptations for excluding, excreting and/or isolating salts within certain tissues, all of which require energy. If the concentration of salts round their roots gets too high, they can have trouble replacing lost water and wilt or even die.

4.3.3 LIGHT

Light can be absorbed by other plants or by turbid water and so limit production on other than a seasonal basis. Fig. 4.2 illustrates a close correlation between

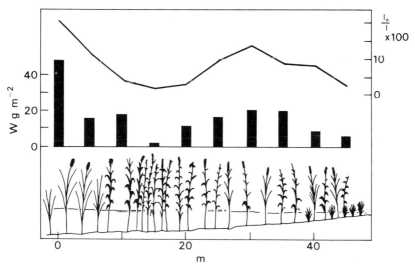

Fig. 4.2. Biomass of duckweeds (columns) growing on the water surface of a Czechoslovakian pond. The line shows radiation flux in relation to full sunlight. The sketches illustrate the structure of the emergent cattail (*Typha*) and reed (*Phragmites*) stands. [After Rejmankova 1975. In: *Limnology of Shallow Waters*, Akadémiai Kiadó, Budapest]

the biomass of duckweed floating on the water surface and the amount of light that can reach them through the reeds which grow up into the air above them. A similar light limitation can be observed by looking at the seasonal changes in production of epibenthic algae growing under salt marsh grasses. During the winter when the grass cover is slight, production of the algae is directly proportional to the strength of sunlight. Production reaches a maximum in

April but then the grasses start to grow and shade the algae so that, during the summer, algal production is limited by low light intensity at the mud surface.

Water lilies with their leaves spread over the water surface have the potential for high competition for light and in dense stands the leaves become partially aerial. Rockweeds are another instance in which the mass of plant material cuts off most light from the underlying fronds. However, at high tide the water support converts the pile of fronds into a stand much like a grass stand in which constant motion allows light to reach all the fronds.

Benthic microalgae are found not only on the surface of sediment but also down into it. Here light is rapidly absorbed by sediment particles. In mud, algae can grow no deeper than 10 to 20 mm before light becomes limiting but in sand with its greater transparency the algae may grow to depths of more than 5 cm.

4.3.4   DEPTH

Fringing plant life rapidly decreases with depth mostly due to the rapid absorption of light by water. However, vascular plants, which have internal gas spaces, or lacunae, are also affected by water pressure. Hydrostatic pressures of less than one atmosphere, equivalent to a depth of 10 m, inhibit the formation of lacunae, and roots. Pressures of only one-half an atmosphere stimulate the elongation of internodes bringing the plant parts closer to the surface. Vascular plants are rarely found growing below about 10 m though the sea grass, *Zostera*, occurs to 30 m off California. The macroscopic algae lack internal gas spaces and are unaffected by pressures within depths where the light intensity is high enough to permit growth.

4.3.5   CARBON DIOXIDE

The CO_2 necessary for photosynthesis can come from the gas in air, dissolved in water, or from dissolved bicarbonate ion. The concentration of CO_2 in air is only about 330 ppm (parts per million). But air is very fluid and easily mixed, which maintains a relatively constant P_{CO_2}, (partial pressure of CO_2,) at the leaf surface. Diffusion of gases in air is rapid so that CO_2 can move rapidly into the leaf to replace that used during photosynthesis. Because the CO_2 level in the air is so low, CO_2 is limiting to photosynthesis for terrestrial plants at high light and nutrient levels and the same may be true at times for emergent aquatic plants. But there are potential compensations for the aquatic plants. They often live in soils very high in organic content in which there is a rapid decomposition, with the evolution of large amounts of CO_2. Most aquatic plants can also use bicarbonate as a carbon source. This is especially important for plants in alkaline waters where the supply of bicarbonate is much greater than that of CO_2. Mosses, on the contrary, are unable to use bicarbonate. For this reason they are usually found only in soft water with low pH. Under these conditions most of the inorganic carbon is present as free CO_2 and available to mosses. This limitation is shared by some algae though most are able to use bicarbonate for photosynthesis.

The lacunae of aquatic vascular plants play an important role in supply of

CO_2 for photosynthesis as well as providing a route for oxygen diffusion to the roots. The roots of these plants are typically growing in sediment with a high P_{CO_2}. CO_2 diffuses from the sediment into the lacunae and can then diffuse throughout the plant. Concentrations of CO_2 in the lacunae are as high as a few per cent compared with the parts per million in the atmosphere, so the lacunal gas can supply a significant portion of the CO_2 necessary for photosynthesis. Those aquatic vascular plants with the C_4 photosynthetic pathway, e.g. the salt marsh *Spartina*, are able to continue photosynthesis down to CO_2 levels of between 0 and 10 ppm compared to the CO_2 compensation point of the C_3 plants which lies between 40 and 70 ppm, making it even less likely that the aquatic C_4 species will be severely limited by CO_2 supply. The lacunae also give the aquatic vascular plants storage space for keeping some of the CO_2 they produce by respiration at night. The lacunae are almost certainly more important for storing photosynthetically produced oxygen for night-time respiration but the mechanism must work both ways and assure the plants of a plentiful CO_2 supply when photosynthesis begins in the morning.

4.3.6 TOXINS

The toxins that might limit plant growth in fringing communities aside from poisons introduced by man, are most commonly salt and H_2S. Salt is a toxin for land plants and so has required an adaptation by plants with terrestrial ancestors that live at the edge of the sea such as the marsh and sea grasses. The marsh grasses are better studied and are known both to exclude salts at the roots and secrete excesses that do get in through special glands in their leaves. Seaweeds have evolved in the sea and share with other algae the ability to regulate their internal ionic composition in relation to sea water.

Plants living in anoxic sediments, especially if the sediment is saturated with sea water, are almost certain to encounter hydrogen sulphide as a result of the reduction of sulphate by bacteria living in the oxygen free environment. Hydrogen sulphide is poisonous but is also readily oxidized to thiosulphate or sulfate before it can reach the root. Occasionally, for reasons not well understood, there are die-outs in salt marshes which may be due to a failure in this mechanism and consequent sulphide poisoning.

4.3.7 NUTRIENTS

Along with water and light, the macronutrients are common limiting factors for production in plant communities. Air that brings CO_2 to the plants cannot carry nutrients. Fluid water can, either as dissolved chemicals or as compounds adsorbed to particles. So the fringing plants, especially those that are partly aerial and partly aquatic, have the best of all possible worlds. Dissolved nutrients are carried by the water that circulates in and out of the fringing communities or flows through them.

Fringing communities are typically depositional areas. Water movements are slowed by the plants and particles suspended in the fluid tend to settle. Sediment from the surrounding bottom stirred by tides or storms, or carried by erosion from the watershed, bring their adsorbed nutrients and deposit them around the roots and holdfasts. The result is that these plants tend to find themselves in an environment rich in nutrients.

Nevertheless, production is often nutrient limited; though high it would be higher yet if all nutrients were present in optimal concentration. It is phosphorus in fresh water, nitrogen in marine and some freshwater systems that have been most often found to be limiting. The micronutrients are less likely to be in short supply, especially in marine systems. These needed elements are all present in sea water and even the macronutrient potassium is present in ocean water in high enough concentration that it does not limit production.

Phosphorus is the most studied nutrient in freshwater systems. Phosphorus from detergents and fertilizers has been the source of trouble in many lakes by stimulating the growth of algal blooms, frequently of the nitrogen-fixing blue-green algae. Phosphorus can also be limiting for productivity of fringing plants, partly because they also are frequently associated with nitrogen-fixing bacteria and are thus relatively independent of nitrogen supply. On the other hand the greater solubility of phosphorus compounds in oxygen-free soils increases their availability in productive marshes with anoxic soils.

In spite of the need for nutrients by fringing plants they are not particularly effective in conserving the supplies of nutrients they do capture. Fig. 4.3 shows the phosphorus content of a cattail (*Typha*) stand at the edge of a lake. The underground stores of the nutrient are obviously insufficient to supply the summer growth of shoots. Sixty per cent of the phosphorus in shoots must have come from root uptake from the sediment during leaf growth. In autumn when

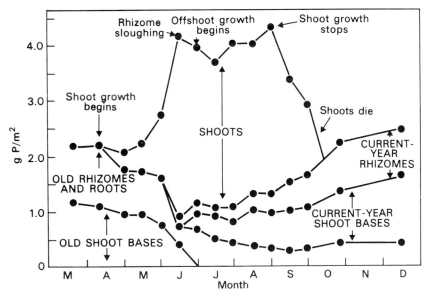

Fig. 4.3. Cumulative seasonal stocks of phosphorus in different parts of *Typha latifolia* plants (as cumulative g P/m²) in University Bay Marsh, Lake Mendota, Wisconsin, 1974. Much of the P in shoots must be from new uptake from the environment since there is insufficient decline in P in the underground parts (old rhizomes, roots and shoot bases) to supply the new growth. Likewise in autumn a major fraction of P loss from shoots does not show up as underground storage for the coming year. [After Prentki *et al.* 1978. In: *Freshwater Wetlands*, Academic Press]

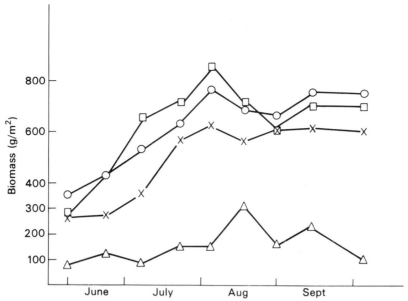

Fig. 4.4. Response of salt marsh to addition of nutrients plotted as living biomass measured above ground. The response to complete fertilizer is not significantly different from that to urea plus phosphorus. The response to urea alone was about 90% of that to complete fertilizer. □ Urea plus phosphate. × Urea. ○ Complete fertilizer. △ Control.

the leaves are dying, relatively little of their phosphorus, less than 25%, is translocated back to the underground rhizomes to serve as a reserve for supplying next year's growth.

Nitrogen is more often limiting than phosphorus for marine fringing communities just as it is for coastal phytoplankton. Fig. 4.4 shows the increases in production in a salt marsh fertilized with complete fertilizer, urea, and phosphate. About 90% of the response to a complete fertilizer was achieved with just the addition of nitrogen (urea). As in the freshwater cattails, only a small fraction of the nitrogen in leaves at the summer's end is translocated to the underground stems.

In fringing communities, vascular plants would seem to have an advantage over non-vascular plants, especially the submersed ones like pond weeds and seagrasses, which can absorb nutrients directly from the water. This is the same nutrient source available to the non-vascular plants. But the vascular plants also have root systems which can absorb nutrients from the pore waters of sediments. Since pore water nutrient concentrations may be more than ten times that of the overlying waters, rooted plants have a large nutrient source available. However, algae living on the surface of the sediment are also fertilized by nutrients diffusing up from the pore waters. Such algae frequently show considerably higher productivity than algae growing on hard substrates which cannot serve as nutrient sources.

In areas where water movements are continuous and bring a steadily renewed nutrient supply the form of the plant makes little difference. Upwelling areas along the coast of California support very large production of kelps. Kelps along the west coast of the Cape Peninsula, South Africa are plentifully supplied with nutrients by upwelling and their production increases as day-length increases in the southern spring. There are pauses in production during the summer associated with temporary cessation of upwelling which indicates nutrient limitation at that time. This contrasts with the situation in Nova Scotia where the kelps grow most actively during the winter—very early spring season when light is minimal but nutrient concentrations are highest.

4.4   Nutrient cycling

Fringing benthic systems at the boundary between terrestrial and aquatic ecosystems, have a considerable influence upon the nutrient supply to natural waters. Because of this activity, they are used in wastewater treatment. The efficiency of nutrient retention by wetlands depends upon how the system cycles the nutrients.

As indicated above wetland vascular plants tend to be rather leaky with regard to their nutrients. They translocate and so save for future growth only a small portion of the nutrients from their above-ground parts. This is true of many plants but for those growing in water, nutrients lost from leaves will not simply fall upon the soil where they may again be captured by the roots of the same or neighbouring plants, but will be carried away, perhaps entirely out of the system, by water flow.

During periods of active growth, what is lost by one plant will be absorbed by another. This can be the case throughout the year in the tropics so that, for example, a riverine marsh in Africa, if water flow is slow, may recycle most of the nutrients that leach from senescent leaves before they have a chance to leave the marsh. At the other extreme, in the autumn in a cold temperate wetland, almost all of the nutrients leached from dead and dying leaves at the end of the growing season will be washed out into the lake, estuary or river that drains the wetland. A tropical rain forest or coral reef maintains its high productivity by conserving its nutrient supply very tightly. Fringing benthic communities depend on their environment for a continual supply.

As an example of a nutrient budget for a fringing community we can look at the data from a tidal salt marsh in Massachusetts, U.S.A. (Table 4.4). The limiting nutrient is nitrogen which is supplied by rainfall, groundwater runoff from the surrounding upland, tidal exchanges from the bay into which the marsh drains, and nitrogen fixation by microbes living in the marsh.

Much of the nitrogen comes with rainfall, either directly upon the marsh surface, which is less than 1% of the total, or on the upland area that subsequently drains into the marsh, about 23% of the total. The nitrogen concentration in the runoff from the land is about twice as great as that in rain because of the loss of water by evapotranspiration before the runoff reaches the marsh. Another major fraction of the supply comes via the activities of the nitrogen fixers living on the marsh surface and in the soil. The fixation due to algae is on the surface of the mud and directly supports the productivity of the mat of blue-green algae but must leach from the algal cells and penetrate into the mud before it can benefit the marsh grasses which are the major primary producers. The bacterial nitrogen fixers on the other hand, are intimately associated with the grass roots and their nitrogen is directly available to the higher plants.

Table 4.4. Nitrogen Budget for a Massachusetts tidal salt marsh. Inputs and outputs are shown as percents. The balance column shows input minus output on an average areal basis.

Category	Input (%)	Output (%)	Balance (kg/ha)
Rain	0.5		+3.9
Runoff	22.5		+152.2
Nitrogen fixation:			
Algal	1.0		+6.3
Bacterial	9.5		+65.2
Tidal exchange:			
Nitrate and ammonia	21.0	30.0	−78.6
Particulate	45.0	49.0	−65.2
Denitrification		18.0	−134.8
Sedimentation		3.0	−24.1
Ammonia volatilization		0.5	−0.4
Shellfish harvest		0.5	−0.4
Gull droppings	0.5		+0.4
			−75.5

By far the largest part of the nitrogen supply comes from tidal exchange. Twenty-one per cent comes as the dissolved inorganic compounds nitrate and ammonia and 45% as particulate matter containing nitrogen. However these same tides are also the most important mechanisms which remove nitrogen from the marsh. In both dissolved and particulate categories, more is carried out of the marsh by the tides than is carried in. The other major loss of nitrogen comes about through the process of denitrification in which nitrate is reduced to nitrogen gas and lost to the atmosphere. The activities of large animals such as gulls and men, the volatilization of ammonia and the accumulation of nitrogen in the sediments are minor corrections to the overall budget which is very close to balancing, a net loss of only 75 kg/ha compared to exchanges of the order of ten times that amount.

The internal cycling in the salt marsh is small in relation to the large exchanges, in contrast to the presumed situation in a tropical forest. A budget for the latter would have large terms for internal cycling such as uptake by roots of nitrogen leached from leaves, and small terms for exchanges with other environments. The marsh acts principally by intercepting nitrogen from runoff and also, during the period of highest production, from the tidal waters flooding the marsh. This nitrogen, with some fixed in place, supports the high production of marsh plants. Some of the intercepted nitrogen is reduced in the anoxic soils and returned to the air but most is exported to the estuary either as ammonia or as particulate organic matter. The total amount of nitrate leaving the marsh is only 40% of that entering from all sources. So the marsh serves as a source of this nutrient to estuarine waters. The amount of particulate nitrogen leaving the marsh is 22% larger than that entering. The marsh serves as a source of particles of organic matter that can serve as food for filter-feeding animals.

One can understand the role of the marsh in nutrient cycling by looking at what would be the exchange between land and estuary if the marsh were not situated between them. The two processes which depend on the organic rich, anoxic sediment—nitrogen fixation and denitrification—would be greatly reduced. Fixation is only about half as great as denitrification, so the lack of these processes would increase the supply of nitrogen to the estuary. The lack of nitrogen uptake by the marsh plants would also result in more nitrogen runoff into the estuary. Since nitrogen is generally the limiting nutrient for phytoplankton production, presumably the lack of a marsh could result in increased eutrophication of coastal waters. The lack of the marsh plants would also do away with the source of particulate matter that comes from the death and partial decomposition of these plants followed by the tidal transport of the resultant detritus to estuarine filter feeders. In short the lack of the fringing marsh system would shift the effect of the runoff from the filter feeding molluscs and detrital feeding worms to the phytoplankton.

There are very few sets of data on nutrient cycling in fringing communities so it is impossible to say whether or not the same holds for other types of fringing systems. But since there are many aspects of these systems which are common, i.e. anoxic soils, high production, and through flow of water, it can be assumed that many of the demonstrated activities of the salt marsh system would also be present in other fringing community types.

4.5 Herbivory as a production control

Herbivores eat plants but, except in the case of phytoplankton, they do not usually eat so much that they limit production. Herbivore populations are normally kept in check by predators. However, when the predators are removed herbivores may increase to such an extent that they drastically limit primary production.

In two separate marine fringing communities normally dominated by kelps, removal of sea urchin predators has resulted in such an increase in urchin populations, that the kelps have been decimated. Urchins are so effective at eating both young algae and the bases by which the older plants are attached, that they can completely wipe out a kelp bed.

The effect of reducing the population of lobsters on the kelp-urchin-lobster ecosystem of Nova Scotia is described on pp. 196–7 and a similar type of relationship exists on the west coast of North America except that the predator controlling the system is the sea otter. Where the otters have been hunted to extinction urchins destroy the kelp and with them go a variety of other animals that depend on kelp for food and shelter.

In freshwater marshes muskrats or nutria, when abundant, can cause what are called 'eat outs'. These are areas where vegetation has been destroyed by both eating and harvesting by the animals for building their lodges. A marsh may have so much of the vegetation removed that appearance changes from a marsh without open water to a lake with only a thin margin of marsh plants. The reduced production under the latter condition plus the influx of muskrat predators such as mink lead to a reduction in muskrats and subsequent regeneration of the marsh. The cycle is complicated on the prairie marshes where it has been best studied by cycles in rainfall, seed longevity, and plant diseases but in general there appears to be a cycle of from 5 to 30 years in which the production varies as the system changes between an emergent plant marsh and an open water pond or lake.

4.6   Summary

Though fringing communities are of enormous variety, they have the potential for very high productivity. They are located where they can intercept nutrients from terrestrial runoff. They have an abundant water supply which is often flowing. Water movements distribute nutrients and remove wastes. The soils are usually anoxic (because they are waterlogged) which favours the activity of nitrogen-fixing bacteria associated with plant roots.

With these combinations of conditions which favour production the fringing benthic communities achieve at their best the highest levels of production known in natural ecosystems.

Further information on these communities may be found in Westlake (1963), Sculthorpe (1967), Wetzel (1975), Chapman (1977) and Goode *et al.* (1978).

5

Detritus and its Role as a Food Source

L.R. POMEROY

5.1   Introduction

Suspended in all natural waters and at the interface of the water with the bottom are assemblages of particles of non-living organic detritus which are derived directly or indirectly from the production of living populations. Associated with these particles are distinctive assemblages of organisms which use the non-living detritus particles as both habitat and a food source. This assemblage of particles and associated flora and fauna is the detritus community, a portion of the total community of natural waters, and a portion of its food web (Fig. 5.1). Although the usual view of the food web is one of grazers cropping plants and in turn being eaten by carnivores, in fact the greater part of plant production is consumed by microorganisms which are eaten by detritivores. In many ecosystems, both terrestrial and aquatic, as much as 90% of primary plant production goes into the detritus food web. The magnitude of the flow of energy through the detritus food web is not generally appreciated. The significance of detritus and the microbial degradation of it as an energy pathway has been minimized by most ecologists, perhaps because we humans are not part of a detritus food web, or possibly because some aspects of it are inconspicuous. The removal of excess plant material makes our planet tidy, but does this contribute to the production of terminal consumers? In aquatic systems, for example, does it feed fishes? Although much, if not most, plant biomass goes into the production of detritus, it has been widely believed to be a dead end so far as the flow of energy is concerned. As a result, detritus and its associated organisms have been largely ignored in studies of food webs in ecosystem, community, and population ecology. However, recent research on the fate of detritus suggests that the detritus food web does interconnect with the grazing food web, and that a significant flow of energy and materials from detritus to terminal consumers does exist.

Because the detritus community has been viewed until recently as peripheral to the main pathways of energy and less research effort has been directed toward it, many gaps remain in our understanding of the detritus food web. In pelagic communities especially, as compared to benthic or terrestrial ones, the detritus food web is difficult to perceive, because most of the detritus particles and associated organisms are quite small and widely dispersed in the water. Water cannot be examined directly, even with the aid of a microscope, in order to see the detritus community. All observations, except those of the kind done by Sheldon *et al.* (1972) with particle counters, require some preliminary concen-

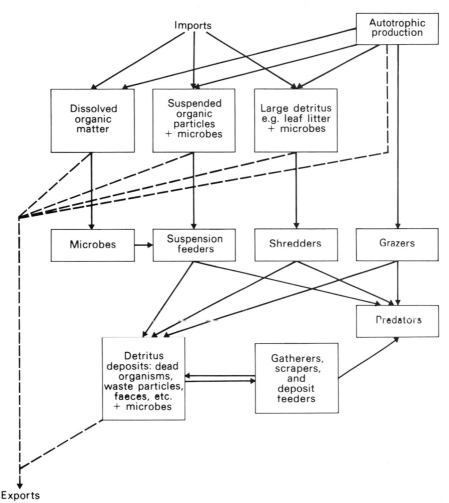

Fig. 5.1. Food resources and their utilization in aquatic ecosystems. [Modified from Berrie A.D. 1976. Detritus, microorganisms, and animals in fresh water. In: J.M. Anderson & A. Macfadyen (eds.), *The Role of Terrestrial and Aquatic Organisms in Decomposition Processes*, p. 334. Blackwell Scientific Publications, Oxford]

trating procedure to make the particles and organisms sufficiently compacted in the water so they can be found at all. Because concentrating methods may also change the physical form of the detritus or alter the rates of processes within the community, our perception of pelagic detritus is limited. Benthic communities are more compact, and some benthic detritivores are macroscopic, but even they are easily disturbed by sampling. It is necessary to use combinations of field and laboratory methods, both observations and experiments, to attempt to piece together the links in the degradation of organic detritus and the transfer of its energy to higher consumer levels.

5.2 The detritus food web

Many variations on the theme of the detritus food web exist, but a number of common elements occur in virtually all of them. Bacteria or fungi appear to be virtually universal as the initial consumers of detritus, even where the food web has been internalized in a rumen or analogous anatomical structure. Because many compounds are refractory, they are ingested and reingested by detritivores which primarily utilize the bacteria growing on the detritus. Multiple, repetitive processing is the rule. Because repeated ingestion would tend to disperse detritus, there are varied mechanisms for scavenging and reaggregating materials down to the size of molecules in solution. Although there is much to be learned, structure can be perceived in the detritus food web. That structure allows for the development of numerous specializations by a diversified community in which interactions between bacteria, protozoans, and metazoans promotes the continuity of the entire detritus community.

Detritus tends to accumulate. In most ecosystems the bulk of organic particulate matter is non-living detritus and only a small part is living organisms. The mass of detritus at any instant is commonly ten to one hundred times that of living organisms. This fact has prompted some scientists to suggest that detritus is not utilized effectively. However, it must be utilized as rapidly as it is produced, or as Wangersky (1977) has stated, we all should be buried in it. There are good reasons why a relatively large standing stock of detritus is logical and beneficial. The detritus community, especially the microbial part of it, lives in a gingerbread house. The food source is also the habitat. If the food source were virtually all consumed, the habitat would disappear. In some aspects of the processing of detritus this does happen, but not to the extent that habitat is totally destroyed. Because some detritus is composed of refractory compounds which are degraded slowly, this leads to the accumulation of a large standing stock under equilibrium conditions of production and degradation.

5.2.1 SOURCES OF DETRITUS

The great bulk of detritus originates, directly or secondarily, from plant biomass. Animal biomass is included, but the production of plant biomass exceeds that of animals by an order of magnitude. Animal faeces, which are largely plant material, are a major secondary source. In larger water bodies most of the detritus is derived from aquatic plants. However, all water bodies including the ocean have a watershed, so in all cases a terrestrial component exists in aquatic detritus. In small, forest-covered streams the terrestrial contribution is 99%, and events in the watershed dominate the stream. In temperate regions the biota 'arranges' its life histories to take advantage of the abundance of leaves in fall and winter. In the tropics the abundance of stream biota is heavily dependent on the presence and species composition of the bordering vegetation, but the seasonal regime is muted.

In water bodies larger than small streams or ponds, the major source of detritus, so far as we know, is vegetation growing in the water, including both rooted or attached macrophytes, algae and higher plants, and the microscopic phytoplankton floating in the water. In lakes with regions of shallow water, in slowly flowing rivers or swampy rivers, in estuaries, and in the coastal zone of the ocean, attached macrophytes usually make the major contribution to detritus, because they are the major producers of plant biomass (p. 71). In fresh waters there are also significant, sometimes dominant, populations of floating plants. Frequently the emergent, floating freshwater plants have very rapid growth and therefore are major producers of detritus. In the North Atlantic Ocean populations of the floating seaweed, *Sargassum*, are impressive, although they contribute relatively little to total plant production.

The process of detritus production from living macrophytes is typically a continuous one. As aquatic grasses grow, leaves or leaf tips die and disintegrate. The same is true of the thalli of kelps and other large seaweeds. The process of detritus formation is accelerated both by the physical forces of waves and the biological processes of grazing. Waves break off branches or entire plants. Grazers are at least as inefficient as other animals, often wasting more than they actually consume, but contributing to the supply of detritus.

In open water and deep water all significant plant production is by phytoplankton (pp. 7–10). Although it has been widely believed that phytoplankton are harvested efficiently by grazing zooplankton, recent studies, such as that of Porter (1973), reveal that zooplankton, too, are messy and sometimes finicky. Chains of diatoms frequently are broken as they are grasped, with a part being damaged and lost. White cells of optimal size are ingested as deftly as so much popcorn, larger cells are crumbled and partially wasted. Moreover, some species of phytoplankton are noxious or even toxic, and they are selectively avoided. Consequently those species grow until depleted of nutrients, die and then are degraded to detritus by microorganisms. We do not yet have accurate measurements of the grazing efficiency of zooplankton under natural conditions, but estimates vary from 90% to 50%. While this is better than the efficiency of grazing on most macrophytes, there is nevertheless a substantial production of detritus directly from phytoplankton (Fig. 5.2).

Throughout the growth and death of aquatic plants, both phytoplankton and macrophytes, there is a direct and continuous loss of dissolved organic compounds (pp. 37 & 69). Some of the early products of photosynthesis, such as glycollate, glucose and other monosaccharides, and amino acids, leak out of living plants. Dying and dead plant tissues rapidly lose soluble materials, and as bacteria and fungi begin to degrade the plant tissues with extracellular enzymes, further losses of soluble compounds occur. But the soluble material is not lost to the detritus food web, since much of it is scavenged by bacteria, either growing free in the water or attached to particles of detritus.

Another major source of detritus is the production of faeces, especially from grazers. From 10 to 50% of the plant materials actually consumed by grazers will not be digested but will be voided as faeces or pseudofaeces. The latter are materials deliberately sorted out and rejected, either during feeding or in the stomach. The faeces, having been partially digested, are a rich source of both dissolved and particulate organic matter, mixed with an inoculum of bacteria. Because many organisms compact or encase faecal matter, the faeces tend to sink relatively rapidly, taking detritus to deeper water or to the bottom. The accumulation of faeces on the bottom probably is a major source of nutrition to benthic communities, especially those in shallow water.

In addition to the concentration of detritus in faecal pellets, other processes produce aggregates of detrital materials. Dissolved organic materials are known to concentrate on the surface film of water and also on the surface film of

Fig. 5.2. Scanning electron micrograph of material filtered from water in a senescent, decaying bloom of *Oscillatoria*. A single chain of dead *Oscillatoria* cells crosses the field. Most bacteria appear to be free individuals which have been collected on the surface of the filter. [Preparation and photograph by T. Jacobsen]

bubbles. If wave action drives a bubble down into the water even a fraction of a metre, the gas will dissolve in the surrounding water, the bubble will collapse, and the organic matter formerly on the surface film becomes aggregated into a microscopic particle. Probably there is rearrangement of chemical bonds at the same time, possibly enhanced by ultraviolet radiation near the water's surface. The result is a flake-like particle 5–25 μm in size (Fig. 5.3). The origin and fate of aggregates is discussed in detail by Riley (1970). Other aggregates may result from the production of adhesive substances by organisms. For example, bacteria produce adhesive secretions to attach themselves to objects and possibly as sites for adsorption of organic matter (Fig. 5.4). Higher organisms produce mucus for a variety of reasons: to capture food or cleanse themselves. These sticky compounds rapidly pick up particulate matter and form flocculent aggregates. Staining to determine the chemical composition of flocculent aggregates reveals that mucus is usually a minor component, although we know that large amounts of mucus are produced, both in the ocean and in shallow waters. Estimates of mucus production have been made on several occasions on coral reefs, and it was found to be equivalent to 2% of total primary production. It is not surprising, therefore, that chemical staining of aggregates collected from the water flowing off coral reefs contained little mucus. In addition to the observation that mucus production is probably equal to only a small percentage of total detritus production, it has been observed (again on coral reefs) that bacterial degradation of mucus is very rapid. Therefore, mucus is not a component which accumulates in detrital aggregates.

88

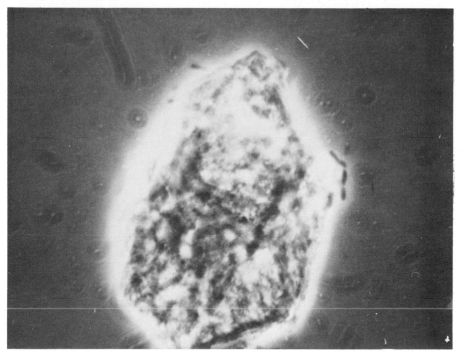

Fig. 5.3. Oil-immersion photomicrograph of a flake-type aggregate. Four actively growing bacteria are seen on the edge of the flake, which is otherwise devoid of bacteria. Vertical dimension of the flake is 25 μm.

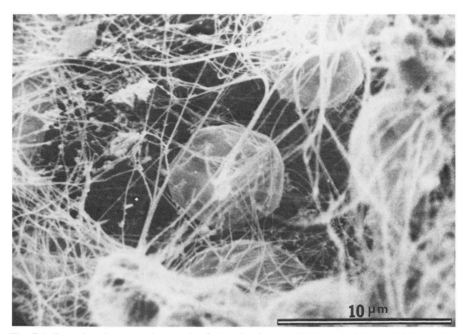

Fig. 5.4. Scanning electron micrograph of strands of extracellular material secreted by marine bacteria. [Preparation and photograph by T. Jacobsen]

Sometimes the aggregation of detritus proceeds until the individual aggregates are visible to the unaided eye. Divers and observers in submersible craft frequently report what they call 'sea snow': aggregates approximately the size of a snowflake and frequently in sufficient density to create the impression of an underwater snowstorm. On at least one occasion, ocean divers reported finding aggregates greater than 10 cm across but so delicate that the aggregates disintegrated as a diver swam past them. A sample of one of those aggregates proved to contain large numbers of both active bacteria and viable single-celled algae. Usual collecting methods, using water samplers which are lowered blindly into the sea, cannot find or collect such objects. Therefore, the perception of aggregates may be biased by the methods employed, and the most active parts of the detritus community may not be collected, because they are at once highly condensed and widely dispersed on a scale which cannot be sampled effectively.

5.2.2   THE DETRITUS COMMUNITY

Several major taxa, or perhaps trophic types, have distinctive roles in the production and consumption of detritus. These include the bacteria and fungi, which are the primary degraders, protozoans which consume bacteria, larger detritivores which consume detritus more or less in bulk, and mucous-net feeders which filter very small detritus particles and even free bacteria from the water. These groups of organisms form a sort of processing line in which some organisms consume particles, some produce particles, and some do both. Details of these processes are discussed by Anderson and Macfadyen (1976) and by Dickinson and Pugh (1974). At the end of the line, the detritus has been largely converted into living biomass again or degraded to carbon dioxide.

The roles of bacteria

The versatile, adaptable bacteria play many roles in the assimilation and degradation of detritus. Their most significant role is the degradation of *everything*. Bacteria, because of their great enzymatic versatility, are able to degrade most naturally produced organic compounds and the majority of synthetic ones. Exceptions include aquatic humus, synthetic polymers, some insecticides and herbicides, surfactants and polychlorinated biphenyls. Only a few of the most refractory plastics, such as polyethylene, appear to be totally resistant to bacterial degradation. To keep our biosphere clean is the traditionally recognized role of bacteria, but that is only the beginning. Because they consume almost everything in their environment, both particulate and dissolved, and because they grow and reproduce more rapidly than other living organisms, the rate of production of bacterial biomass is potentially high (see below). This means that bacteria can be a major source of energy for other consumers. Even though the standing stock of bacterial biomass in an aquatic ecosystem may never be as great as the standing stock of phytoplankton or macrophytes, the bacterial contribution to secondary production can be substantial, because the bacterial biomass may be replaced several times each day under optimal conditions (Fig. 5.5) (Sieburth 1976; Watson *et al.* 1977). Conditions will be

90

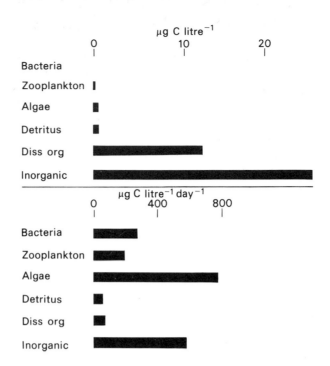

Fig. 5.5. Structure of living and non-living organic materials in Frains Lake, Michigan. Upper bar graph shows initial concentration of components at sunrise. Lower bar graph shows rate of production of the components. [From Saunders G.W. (1972). The transformation of artificial detritus in lake water. In: U. Melchiorri-Santolini & Hopton J.W. *Detritus and its Role in Aquatic Ecosystems, Mem. Ist. Ital. Idrobiol.* **29,** Suppl. p. 280]

optimal when there is enough organic matter to support an exponential rate of growth and when other environmental conditions are acceptable.

We know that under laboratory conditions, when microorganisms are supplied with all requirements for growth, production of microbial biomass is very rapid. However, it is difficult to transfer that information to the real world, because conditions there may be less than optimal most of the time. To understand the production of microorganisms, it would be informative to have measurements of their growth rate in nature or as nearly so as possible. That has proven difficult. We cannot simply count the microorganisms in a water body from time to time, because they are being removed by grazers and filter feeders. However, microorganisms can be fairly well separated from their grazers by passing water gently through a very fine screen, usually one with a porosity of 10 μm or less. The filtrate is then held in a bottle or a dialysis chamber, and the change in number of microorganisms or of some parameters of microbial biomass is measured over time. Some investigators make direct counts of bacteria. Others measure uptake of bicarbonate or an organic carbon source labelled with carbon-14. It is also possible to measure microbial respiration and to relate it to growth rate by making an assumption about assimilation efficiency. While direct counts measure production of bacteria only, some of the indirect methods potentially include production of fungi and the small blue-green algae, the Cyanobacteria. Obviously, all of these methods have shortcomings; each has a different set of them. Comparisons of the methods have shown that at least in some circumstances there is surprisingly good agreement, but this is not true in all cases. Compared with the vast quantities of data on production of phytoplankton, and even with the data on secondary production of higher organisms, the data on production of microorganisms are sparse indeed.

Moreover, because of the lack of one generally accepted method, and the diverse problems associated with current methods, the few existing data cannot be closely compared. At best, we can distinguish rates differing by an order of magnitude. However, even that is useful, because the rates of microbial production span several orders of magnitude. Representative data on the production of microorganisms in the water column have been assembled in Table 5.1. Generally the absolute rates of microbial production are high in eutrophic waters, such as Dalnee Lake or coastal upwellings, and low in oligotrophic waters, such as the tropical Pacific or Lake Biwa. However, some investigators believe that the production of microorganisms, relative to the secondary production of macroscopic grazing organisms, is greater in some oligotrophic systems, such as the central oceanic gyres.

In shallow water bodies, where most macrophyte production becomes detritus and where there may be substantial input from the land as well, the production of microorganisms sometimes exceeds the production of phytoplankton. This is not surprising, when one considers the sources of organic matter available to microorganisms in such systems. However, several investigators have reported rates of production of microorganisms in the open ocean which exceed the accepted rates of phytoplankton production. Since there are no known sources of primary production of any significance in the open ocean except for phytoplankton, this appears to leave the microorganisms without an adequate source of energy. Obviously, there is a problem in estimating primary production, microbial production, or both in the ocean. If bacteria are indeed consuming much of the primary photosynthate, they are either an important link in the food web or a major sink for energy. These questions remain to be answered.

Production of microorganisms in the bottom sediments of water bodies varies with primary productivity and with depth of the water. Very shallow water bodies have more benthic than pelagic microbial production, whether they are eutrophic or oligotrophic. For example, the arctic pond in Table 5.1 had a daily benthic microbial production rate of $417 \, mg \, C \, m^{-2} \, day^{-1}$, assuming an active season of 60 days. If the water column is long, however, most secondary production occurs as the detritus is sinking, and relatively little organic matter reaches bottom. That is why bottom sediments and bottom waters of ocean basins and continental shelf seas are almost never anaerobic. Only the Black Sea among the larger continental seas is anaerobic at the bottom, owing to stratification caused by the influx of fresh water at the surface.

While we have many more questions than answers regarding the rate of production of microorganisms in natural waters, there is reason to believe that we are on the threshold of perceiving this portion of the aquatic food web in considerable detail. Methods now at our disposal make it possible, and enough is known to suggest that this endeavour is worth our attention.

The conversion of detritus into bacterial or fungal biomass makes it much more readily available to other organisms. Some bacteria clearly are resistant to digestion, but most are not. Bacteria, whether single and floating free in the water or attached to detritus particles, can be consumed and digested by larger organisms which lack the digestive enzymes necessary to degrade some of the principal components of plant detritus, such as cellulose. Because of their high

Table 5.1. Estimates of microbial production rates in natural waters. Means were calculated and in some cases terms of reference were transformed for uniformity.

Location	Production rate $(mg\,C\,m^{-3}\,day^{-1})$	Production rate $(mg\,C\,m^{-2}\,day^{-1})$	Method	Source
Arctic pond		5*	direct count	Hobbie and Rublee (1975)
Lake Biwa, Japan	11.5		dark HCO_3	Mori and Yamamoto (1975)
Lake Suwa, Japan, summer		900	direct count	Mori and Yamamoto (1975)
Lake Suwa, Japan, winter		100	direct count	Mori and Yamamoto (1975)
Frains Lake, Michigan, U.S.A.	300		respiration	Saunders (1972)
Dalnee Lake, Kamchatka	1530		respiration	Sorokin (1978)
Ryblinsk Reservoir, U.S.S.R.	57	457	dark HCO_3	Kusnetsov and Romanenko (1966)
Narragansett Bay, U.S.A.	202		direct count	Sieburth et al. (1977)
Butaritari Atoll Lagoon		410	respiration	Sorokin (1978)
Continental shelf water behind Great Barrier Reef		810	respiration	Sorokin (1978)
Black Sea, euphotic zone	2		respiration	Sorokin (1978)
Black Sea, chemosynthesis zone	6		respiration	Sorokin (1978)
Peru coastal upwelling	11	1600	respiration	Sorokin (1978)
N. Atlantic E. of Azores	120		direct count	Sieburth et al. (1977)
Sea of Japan	45		respiration	Sorokin (1978)
N. & S. Pacific, mid and low latitudes	0.6	126	respiration	Sorokin (1978)

*Assuming 60 day active season.

surface to volume ratio, bacteria are efficient scavengers of dilute dissolved organic compounds. They are able to utilize them in natural waters down to concentrations of a few micrograms per litre. While many higher organisms have the demonstrated ability to absorb dissolved organic materials from solution, they are generally less efficient than bacteria, and in many cases they are losing more dissolved organic matter than they are absorbing, given the very low concentrations which are the usual case in natural waters.

Roles of detritivores

Free-living Protozoa are mostly specialized detritivores which do not ingest whole detritus particles but graze among them, consuming the bacteria growing on the surfaces of the particles. Except for the larger ciliates, most protozoans are in the same size range as the detritus within which they feed, so they will graze the surfaces or burrow through flocculent aggregates, seeking and consuming individual bacteria. Some of them also filter volumes of water, extracting individual free-living bacteria from it.

Protozoa are perhaps the most ubiquitous but the least studied ecologically of the detritivores. Wherever bacteria begin to multiply, protozoans soon appear, and bacterial numbers diminish forthwith. By keeping bacterial populations sparse the protozoans serve to keep them in an exponential growth state, promoting bacterial production and the conversion of detritus to living biomass. The interaction of the protozoans with their bacterial food source is analogous to growing bacteria in a continuous culture, and models of this interaction, such as that of Hunt et al. (1977), have been developed. In continuous culture in the laboratory, bacteria are supplied continuously with fresh nutrient media while at the same time a part of the culture is drawn off at a rate which just balances the rate of growth of the bacteria. Thus an optimal and continuous production of bacteria is achieved. Protozoans in nature will not be able to adjust perfectly their grazing rate to the supply of bacteria, and the supply of new nutrients (detritus) will be more or less irregular, but the effect of grazing by Protozoa is analogous to the continuous culture and it will extend significantly the period of exponential growth of the bacteria.

Circumstantial evidence that bacteria are constantly grazed by protozoans or other detritivores is found in observations of freshly concentrated samples of the living detritus community. When first concentrated and observed under the microscope, most detritus particles appear to be virtually devoid of bacteria (Figs. 5.1 & 5.3). Relatively more bacteria are seen floating free in the water than attached to particles. However, if the concentrates are kept aerated and are held for 12 to 24 hours, bacteria appear on the surfaces of the flake-like particles and embedded in the flocculent aggregates (Figs. 5.6 and 5.7). They divide and begin to form colonies. The process of concentration isolates the bacteria from at least some part of the populations of organisms which consume them, and their rate of growth during the first 24 hours of isolation may be at least indicative of bacterial production rate.

The same relationship may exist between ciliates and populations of bacteria which live as free individuals in the water, existing on dissolved organic compounds. Not all filter-feeding organisms can remove from the water particles as

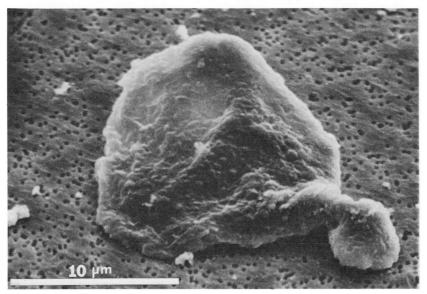

Fig. 5.6. A small, flake-type aggregate collected on a 0.2 μm membrane filter. [Preparation and photograph by T. Jacobsen]

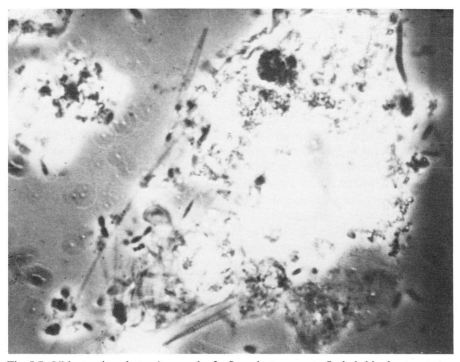

Fig. 5.7. Oil-immersion photomicrograph of a flocculent aggregate. Included in the aggregate are several actively growing bacteria. Width of the field is 30 μm.

95

small as a single bacterium, of the order of 1 μm, and this may give the free living bacteria some refuge from grazing. However, the ciliates can filter and consume individual bacteria. Because of their small size and relatively low numbers (10^3–10^4 per litre), ciliates are difficult to discern in samples of natural waters. The samples must be concentrated in such a way that ciliates are retained but not destroyed by the procedure. When this is done, ciliates usually are found to be present in numbers sufficient to have a significant impact on bacterial populations.

A diverse group of detritivorous macrofauna, principally invertebrates, play a role in the detritus food web. These particularly include holothurians, echinoids, and polychaetes in the sea, and insect larvae and oligochaetes in fresh waters. They consume detritus and presumably they digest the bacteria (Lopez & Levinton 1978), but pass most of the detritus itself as faeces. In the process of doing this they may concentrate the detritus into larger units (faecal pellets), they may separate organic detritus from inorganic materials such as sand, and they may grind up individual detritus particles into smaller ones.

The shredding and grinding role of detritivores has been given detailed attention in streams, where specialized guilds* of insect larvae process leaves, stripping the leaf blades off the petioles and grinding them into small pieces. Probably because of the efficacy of these stream insects, leaves of terrestrial origin cease to be a major component of the detritus food web in most larger streams and rivers. However, other materials receive similar treatment by other groups of invertebrates in many aquatic environments. Even at the bottom of the ocean, shredding by invertebrates appears to be an essential step in the degradation of organic matter. Some serendipitous events which followed the sinking of the submersible research vessel *Alvin* led to this discovery.

During a launch of *Alvin* from its mother ship, *Lulu*, a wave filled the vessel, and it sank in 1540 m of water. The crew escaped, but in their haste they left their lunches aboard. When *Alvin* was raised to the surface 10 months later, scientists on the recovery vessel were surprised to find the lunches well preserved. Most of the food was edible, albeit soggy. Food kept in the refrigerator for a similar time at the same temperature would have been less well preserved. Was the slow rate of degradation the result of a combination of a low temperature and high pressure? Perhaps, but the observation led to a number of experiments by several investigators. Various defined organic materials, as well as some portions of organisms, were placed on the sea bottom for extended periods to observe the rate of degradation. In one series of experiments some of the materials were protected by multiple wrapping, like sandwiches in a lunch box, while others were exposed completely in coarsely screened cages. Those materials which were wrapped would become saturated with sea water containing the natural microbial flora while those in the cages would be exposed to the attack of small invertebrates as well. It was found that the exposed materials were degraded quickly, after invertebrates had shredded them, while the protected materials were not. Similar experiments in forest litter communities have produced comparable results.

When leaves of either terrestrial or aquatic origin are allowed to degrade

*A *guild* is a group of species having similar niches and performing similar ecological roles.

96

under laboratory conditions, the rate of degradation is often quite slow. Finely-ground eelgrass detritus showed a loss of 12% of organic matter if held with added bacteria. In such experiments shredder organisms are not included, and it appears that the shredders may do something which cannot be duplicated simply by grinding up the leaves. In some natural systems there is also a lack of shredders, and leaves tend to persist in the water column or at the surface of large water bodies, for example. Probably the supply of leaves is sparse or intermittent, so no guild of specialists in shredding can survive. In the absence of a shredding guild, detritus which requires it to become more generally available may at times accumulate. We frequently see leaves floating on large water bodies, including the ocean, at locations which indicate that the leaves have been afloat for weeks or even months.

Equally important, especially for the degradation of some of the more refractory compounds, is the packaging of detritus into faecal pellets (Fig. 5.8).

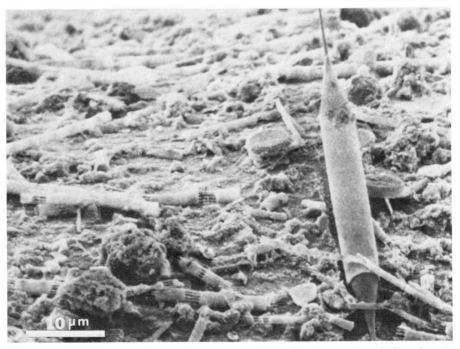

Fig. 5.8. Scanning electron micrograph of estuarine detritus from Doboy Sound, Georgia, U.S.A., collected on a 0.2 μm membrane filter. The diatoms, *Skeletonema*, and *Rhizosolenia*, are present. Most of the detritus appears to be the remains of disintegrated faecal pellets. [Preparation and photograph by T. Jacobsen]

It is relatively easy to observe the fate of a faecal pellet, and this has been done many times, with quite a few species of invertebrates. The crustaceans, such as planktonic copepods or epibenthic shrimps, produce faecal pellets encased in a tube of chitin. The pellets remain intact for several days in some instances, but because of their small size and transparent casing, events inside the pellet can be observed under a microscope. With the aid of the DNA-binding fluorescent stain, acridine orange, it is possible to observe the rapid multiplication of bacteria inside the pellet during the first day. Usually by the second day there are many motile bacteria, probably indicative of reduced concentration of dissolved oxygen. By the second day protozoans also appear and increase in numbers. During this time it is not unusual to observe phytoplankton cells which have survived passage through the gut and which appear to be intact and in good health. Given sufficient light, presumably they would grow in the nutrient-rich pellet. By the end of the second or third day of observation, chitinoclastic bacteria degrade the casing, and the pellet disintegrates, leaving scattered small detritus particles and microorganisms which await another filtration and concentration into a faecal pellet.

Because they are larger and heavier than the individual detritus particles, faecal pellets fall to the bottom, and the larger ones which fall quickly are a major source of energy for benthic detritivores. Within benthic communities several modes of feeding on detritus occur. As in the plankton, the protozoans and some of the meiofauna graze on individual bacteria, fungi, and whatever else they can. Other organisms are quite selective, doing considerable sorting as they eat. Further sorting often is accomplished in the stomach, where sand and other large particles are separated from fine organic detritus. The large material is formed into faeces at once, while the fine material is shunted into digestive diverticula. Reports of experiments with both benthic and planktonic invertebrates suggest that some species are able to digest ingested detritus and not merely the bacteria on it. It remains to be demonstrated whether the invertebrates possess the necessary enzymes themselves or whether the bacterial flora in their guts is providing the digestive expertise in exchange for the usual energy cost of bacterial digestion and assimilation.

Not all faecal pellets are large, heavy, and durable. A substantial fraction of faecal material is not compacted or is sufficiently small to remain in suspension in the water column until it is degraded. While faecal pellets are an important food source for the benthos, they are also important constituents of the planktonic detritus food web. Another structure which may contribute to the supply of planktonic detritus is the so-called house constructed by the Appendicularia, a group of pelagic tunicates. The individual appendicularian is inside the house and pumps water through it, filtering out nannoplankton and probably bacteria. Particles larger than about 25 μm cannot be ingested by the animal, and in time the larger particles clog the filtering apparatus and the house is abandoned. Actively feeding appendicularians may produce and discard several houses per day. The houses then become detritus particles, suitable in size to be filtered from the water by other organisms, and composed of materials which can be degraded readily by microorganisms.

5.3   Cycling of essential elements

All living organisms are composed of a similar suite of chemical elements. If populations of organisms are to reproduce, grow, or even survive and maintain a constant biomass, there must be a continuing supply of all of the chemical elements which are essential constituents of protoplasm. It has been known for a century that if the supply of any one essential element is depleted, this will limit the growth of a population (see pp. 32–37). The supply of elements also limits the growth, and shapes the structure of entire communities. The supply of essential elements to populations and communities may come from outside the system, such as terrestrial runoff into streams, stream input into lakes and estuaries, river input into the ocean, or vertical advection of deep water to the surface of the ocean. However, much of the day-to-day supply of essential elements is derived from recycling within the system. Because much of the flow of energy and materials is through the detritus food web, the supply of elements through recycling is intimately involved with the fate of detritus. Conversely, populations of denitrifying bacteria in the anaerobic sediments of estuaries and other coastal waters are major sinks for available nitrogen.

The traditional role assigned to bacteria, other than that of cleaning up the biosphere, is the regeneration of basic plant nutrient elements, such as phosphorus and nitrogen. The experimental basis for this comes from laboratory studies, either with pure cultures in media containing excess nutrients or in hay infusions and the like. Under these conditions, bacteria release excess phosphate and ammonia into the water. However, tracer experiments under more natural conditions generally show the bacteria retaining all or most of the nutrient elements from the plant materials they degrade. The reason for this can be found in the relative proportions of nitrogen and phosphorus in plants and bacteria. Phosphorus makes up less than 1% of the dry weight of most plants but it makes up about 5% of the dry weight of most bacteria. If bacteria oxidize about half of the plant material they consume as an energy source, but save all of the phosphorus from it, they still require additional phosphorus to sustain growth. The same is true of nitrogen, although the relative proportions are more variable. Therefore, it would be prudent for bacteria to retain as much nitrogen and phosphorus as possible from the detritus they degrade, since they must compete with phytoplankton for additional phosphate and ammonia in solution in the water. If bacteria retain most of the nitrogen and phosphorus available to them, it will be regenerated only when the bacteria are consumed by protozoans or other detritivores, which have a phosphorus content of around 1%. The experimental evidence for the two possible routes of regeneration, directly by bacteria or by way of excretion of bacteriovores, is quite mixed, and there is not general agreement about the role of bacteria and higher heterotrophic organisms in the regeneration of nitrogen and phosphorus in aquatic ecosystems.

No matter which way it works in nature, the cycle of regeneration of essential elements is completed very rapidly, minimizing the amounts of essential elements tied up in non-living organic material. No one doubts that bacteria do indeed perform their role as regenerators of plant nutrients one way or another, or that they do it swiftly. In some lake waters, the residence time of phosphate is only a matter of minutes. That is, the standing stock of phosphate is taken up by the phytoplankton, or bacteria, and replaced by an equal amount every few minutes either by bacteria or through excretion.

Since most of the standing stock in the detritus community usually consists of detritus yet to be degraded, and only a small fraction is living microbial biomass, the proportions of carbon, nitrogen, and phosphorus in detritus do not change greatly with time in most natural waters. Detritus from great depths in the ocean has approximately the same $C:N:P$ ratio as that near the surface (Fig. 5.9). However, if microorganisms accumulate, either on the surfaces of

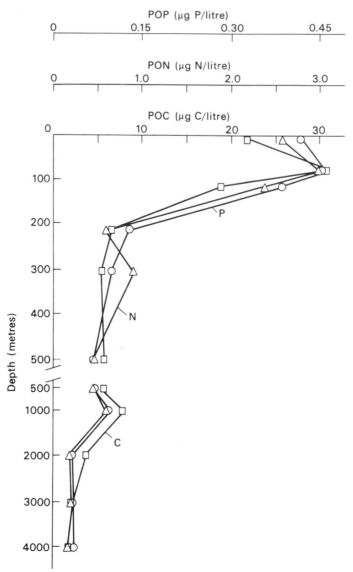

Fig. 5.9. Vertical distribution of total particulate organic carbon, nitrogen, and phosphorus off Baja California, 29° 00′N 122° 30′W. [From Holm-Hansen O. 1972. The distribution and chemical composition of particulate material in marine and fresh waters. In: U. Melchiorri-Santolini & Hopton J.W. *Detritus and its Role in Aquatic Ecosystems, Mem. Ist. Ital. Idrobiol.* **29,** Suppl. p. 280]

detritus or within it where they are less vulnerable to protozoans, the nitrogen content of the detritus increases. A majority of studies of the change in per cent nitrogen of detritus with time show some increase (Fig. 5.10). Terrestrial leaves show the least change. Where an increase in nitrogen is seen, as by Odum and de la Cruz (1967), it indicates growth of microorganisms on or in the detritus, and further suggests that the microorganism have accumulated nitrogen either from the water or from detritus they have degraded.

100

Fig. 5.10. Changes in nitrogen as percentage of remaining dry weight, during the decomposition of detritus derived from aquatic and terrestrial macrophytes. Key: 1, Mangrove leaves in field; 2, marsh grass in field; 3, eelgrass leaves in field; 4, eelgrass leaves in lab; 5–8, eelgrass leaves in lab; 9–17, terrestrial leaf litter decomposed 6 months on forest floor: 9, ash; 10, oak; 11, hazel; 12, alder; 13, chestnut; 14, beech; 15, maple; 16, beech; 17, birch; 18–21, terrestrial leaf litter decomposed in fresh water for 6 months (18) or 50 days (19–21); 18, beech; 19, elm; 20, alder; 21, oak. Dashed line indicates no change in per cent nitrogen. [From Harrison P.G. & Mann K.H. 1975. Detritus formation from eelgrass (*Zostera marina* L.): the relative effects of fragmentation, leaching, and decay. *Limnol. Oceanogr.* **20**, 924–934]

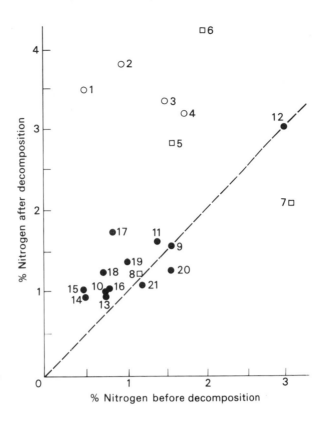

5.4   Efficiency

One of the reasons so few investigators have studied the detritus food web is the widely-held belief that any food web with so many transfers of energy would be highly inefficient and could not possibly contribute a significant flow of energy to the terminal consumers in the ecosystem. That belief is based on the supposition that each transfer should be considered a trophic level, and that the so-called ecological efficiency of energy transfer between trophic levels is about 10%. If this were indeed true, then 100 kcal of plant production, passed through bacteria, protozoans, and macrodetritivores, would yield only 0.01 kcal of terminal consumer biomass. In fact, trophic levels were conceived with the grazer food chain in mind. Even in that case, we now know that the efficiency of transfer of energy between producers and consumers varies from less than 10% to more than 40%. The trophic level concept, applied to the detritus food web, puts bacteria on several trophic levels, equivalent to grazers and even predators. Moreover, most filter feeding organisms are omnivorous. They concentrate and consume whatever happens to be available within the size range of particles they are able to ingest. They will consume phytoplankton, protozoans, or detritus and bacteria. Therefore, they cannot be assigned to one well-defined trophic level. The most helpful approach appears to be the concept of gross growth efficiency (growth/ingestion) as the measure of energy transfer between each producer and consumer, treating the units in the detritus food web as secondary producers and consumers. The gross growth efficiency of the specific kinds of organisms involved—bacteria, fungi, protozoans and invertebrates—is known to some extent. Laboratory studies with bacteria commonly reveal efficiencies in excess of 50%. Unfortunately, we know little about bacterial growth efficiency under natural conditions. Since bacteria have the ability to turn on their metabolic machinery, run until the supplies are gone, and then shut down, they may run intermittently at quite high efficiencies. We are not certain that they always do this in nature. Protozoans also are fast living although not quite as efficient as bacteria. The macroinvertebrates, based now on quite substantial data, appear to have gross growth efficiencies in the 20–30% range. Taking these data, mostly from the laboratory, as assumptions, 100 kcal of plant energy might yield 1–3 kcal of terminal consumer biomass. That would be a viable food web. However, there are insufficient data on conditions in natural ecosystems to test these assumptions. Nevertheless, the efficacy of the detritus food web in providing energy to terminal consumers cannot be found wanting on purely theoretical grounds.

5.5 Community stability

Several features of the detritus food web contribute to the stability of the community and also to its diversity. One is the varying amount of time required to degrade the various plant tissues entering an ecosystem. Low molecular weight compounds are degraded in a matter of days, and most of the compounds with a good balance of essential elements probably are utilized within a few weeks. The refractory materials and those which require shredding accumulate a bit more, providing habitat and at the same time providing a continuing yield of energy to the system between pulses of the more readily degraded materials. A single plant species may provide leaves, stems and rhizomes, each with distinct properties. Each additional species provides another set. The varying decomposition rates not only lead to a relatively steady supply of energy to the consumers but they also provide more niches for specialized organisms to fill.

Considering not only the variety of plants which contribute to the detritus pool, from phytoplankton to trees, but the variety of transformations this material may undergo, including dissolved materials, aggregates, faecal pellets, single bacteria, and fungal hyphae in particles, the detritus food web exhibits at least as much diversity as does the grazer food web. Two extreme examples of purely detrital systems are small streams and the deep-sea benthos, both of which are heterotrophic systems, dependent totally on imported organic matter. In a typical temperate stream there are some dozens of species, mostly of larval insects, which are specialized to process and to utilize in different ways detritus and its associated flora of bacteria and fungi. At the bottom of the ocean, several thousand metres from the nearest functioning photoautotroph, there live hundreds of species, most of which are specialized in minutely different ways to gather, process, and utilize detritus. On the sea bottom, where the detritus food web is the only way of life other than carnivory, one of the most diverse and apparently one of the most stable communities in the biosphere exists. Although the biomass per unit area is small, and the metabolic rate of the community is one of the lowest known, the many species survive and coexist on the rain of detritus from the sun-lit waters above.

102

6

Benthic Secondary Production

K.H. MANN

6.1 Introduction

On an area basis, by far the greater part of the aquatic benthos is found at depths where the light intensity does not permit plants to live by photosynthesis. Under these conditions the benthic community relies for its food on that which sinks from the euphotic zone above. There are phytoplankton cells, faecal pellets and dead bodies of planktonic animals, and if the shore is not too far away there may be remains of large plants such as seaweeds, seagrasses, pondweeds or even terrestrial plants.

For a benthic subsystem of this type, its role in the ecosystem is clear. It receives organic matter from above, passes it through a complex web of macroscopic and microscopic organisms and returns to the system two products: secondary production in the form of invertebrates which can be removed by predators; and a supply of nutrients which have been released from the food as it was processed through the food web. Of the two, the nutrients are the most crucial to the continued functioning of the system. Unless these nutrients are released into the water column and carried by water currents back to the euphotic zone, the downward rain of organic matter will eventually deplete the upper layers of almost all nutrients, so that primary production is drastically reduced. However, benthic secondary production supports a very large proportion of the world's commercial fisheries. Studies of benthic productivity are essential if we are to understand and manage these fisheries effectively.

In shallow waters the benthic animals may be living side by side with the plants on which they depend. Under these circumstances the inputs and outputs of the benthic secondary producers are more difficult to analyse. However, these are the conditions we find in most rivers, in parts of estuaries, and in the littoral zone of lakes and the sea. In what follows, attempts will be made to refer to these situations, even though their total area is small compared with the huge expanses of deep benthos.

The questions we shall address in this chapter include the following: What kinds of organisms make up the benthic secondary producers? What amounts of food reach the benthos in various situations, and how do the organisms exploit it? How stable are benthic communities? How efficient are they at converting food input to secondary production? What are the mechanisms and patterns of nutrient regeneration?

6.2   The supply of food to the benthos

Pictures obtained by television cameras lowered to great depth have shown that on the floor of the deep sea there may be an occasional bounty in the form of a corpse of a fish, or a more or less complete kelp plant, and that when this occurs there are mobile consumers that quickly converge on the feast. Most of the time, however, the food supply is in the form of small particles of dead organic matter, richly coated with bacteria and sometimes fungi. When considering the magnitude of this supply, it is obvious that it must be less than the amount of net primary production at the surface, and that the difference must be caused by removal by consumers in the water column. If the water is deep, there is a high probability that a large proportion of the primary production will be consumed either by animals or by bacteria in the water column. Zooplankton produce faecal pellets, but even these may be reingested by other organisms, or consumed by bacteria as they sink. It has been estimated, for example, that in the Sargasso Sea where primary production is not very high and where the water column is several thousand metres deep a very small proportion, of the order of 1% of surface primary production, reaches the bottom.

Stratified water bodies are divisible into an upper mixed layer in which the wind causes frequent vertical mixing and a fairly uniform temperature, and a lower relatively stagnant, cool layer (pp. 14–16). In the mixed layer phytoplankton is constantly being brought to the surface, and it is kept actively growing so long as there is a supply of nutrients. However, the longer a particular cell stays in the surface waters, the greater are its chances of being consumed by a herbivore. Hargrave (1973) showed that the proportion of surface productivity reaching the bottom is inversely proportional to depth (Table 6.1) but is also

Table 6.1. Proportion of primary production reaching sediment surface in various aquatic ecosystems. [After Hargrave 1973]

Location	Depth (m)	% reaching sediment surface
Lake Washington	60	29
Marion Lake	5	31
Sargasso Sea	4000	0.5
Black Sea	2200	10–15
Chesapeake Bay	5	55
	15	20
Lake Esrom	20	44
Lawrence Lake	12	64

inversely proportional to the depth of the mixed layer. His model (Fig. 6.1) adequately represented all the data he had collected. The data include both marine and freshwater habitats. In general, water over very deep basins tends to be less productive than in shallow basins, and in the very deep basins the proportion reaching the bottom is low. This means that there is a very large spread in the values for food input to the benthos. In midocean areas with surface productivity at perhaps $50 \, \text{g C m}^{-2} \, \text{yr}^{-1}$ and only 1 or 2% of the

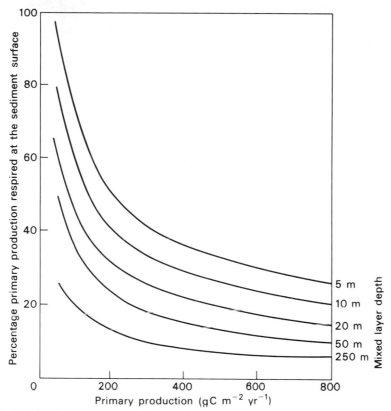

Fig. 6.1. Plot showing percentage of surface primary production respired at the sediment surface. [From model relationship of Hargrave 1973]

production reaching the bottom, the annual input may be less than $1 \text{ g C m}^{-2} \text{ yr}^{-1}$. At the other end of the scale we have eutrophic lakes with productivity in excess of $500 \text{ g C m}^{-2} \text{ yr}^{-1}$ and a mixed layer depth of less than 5 m, where 200 g C m^{-2} may reach the bottom each year.

Benthic communities in shallow water commonly include macrophytes such as algae, or various aquatic flowering plants (see p. 67). It is characteristic of aquatic macrophytes that they are seldom consumed while living, but die and decay, to be utilized by detritivores. The most usual habitats for such plants are rivers, estuaries, or the wave-washed shores of lakes and the sea. All are characterized by fairly vigorous water movement. It is not uncommon for beds of macrophytes to produce $1000–2000 \text{ g C m}^{-2} \text{ yr}^{-1}$, but it is not usual for the benthic animals living in the same area to consume all this production. Instead, it is dispersed by water movement over a wide area, and the local benthic community probably consumes no more than 10% of it.

6.3 The organisms of the benthos

The conventional way of investigating and discussing the role of the benthos is according to the size of organism involved. Organisms over 1 mm, which are

105

normally invertebrates, are referred to as the macrofauna, or macrobenthos. Those below 0.1 mm are the microbenthos, and those in the middle range are the meiobenthos. We have a great deal of information about the kinds and numbers of animals in the macrobenthos, and their way of life. Work on meio- and microbenthos is still in its infancy. A common sampling method is to take a bite with a grab, and sieve out those organisms living buried in the sediments. The most common kinds of these infaunal animals in the sea are annelids and bivalve molluscs, while in lake and river sediments the larvae of chronomid flies can also be very abundant.

Common epifaunal animals include mobile forms such as shrimps, crabs and lobsters, gastropods, starfish and sea urchins, and sedentary forms such as barnacles, sea anemones, sponges and bryozoans which are attached to hard surfaces. There is a considerable group of animals which burrows during the day, emerging at night to feed; so the division of infauna and epifauna is not a rigid one.

6.3.1 FEEDING MECHANISMS

If we think for a moment about the rain of food to the benthos, there is competitive advantage to be obtained from filtering the food material from the water before it even reaches the bottom. Hence the abundance of filter-feeding organisms. Many animals, such as certain polychaete worms, chironomid larvae, and some crustaceans, make U-shaped burrows in the sediment, pump a current of water through it, and filter the food particles. Others, such as bivalves, and ascidians, have inhalent and exhalent siphons and a filtering mechanism between the two.

Some polychaete worms extend a 'fan' of tentacles to catch the rain of sediment, withdrawing rapidly into a tube at the approach of a predator. Large areas of the sea floor are dominated by such filter feeders. The reasons are not entirely clear, but they may be places where the water movement keeps the organic matter in suspension, not allowing very much to settle on the bottom. In fast-moving rivers this is the case, and there is often an abundance of animals which simply filter the passing current. Examples include black-fly larvae and net-spinning caddis-fly larvae.

In areas where the food supply settles on the sediment surface, there is an assortment of infauna and epifauna to exploit it. Many worms, holothurians and some bivalves simply ingest the sediment in which they burrow, but others have devices for collecting food from the sediment-water interface. Gastropods and amphipods may browse on the organic debris, while shrimps cruise over the surface and select choice morsels.

One of the end results of all this feeding is that faecal pellets are formed and these are almost always deposited at the surface. Many organisms are happy to ingest faecal pellets, and it has been shown that each time faecal material passes through the gut of an animal it is stripped of its coating of microorganisms (see p. 90). As it lies exposed to the environment it is again attacked by a microflora, which is again removed by the next animal to consume it. In this way plant material such as cellulose or lignin, which may resist digestion by invertebrates, is progressively degraded. The constant activities of macrobenthos, in burrowing and depositing faecal pellets, results in a vigorous turning over of the first few centimetres of sediment, a process known as bioturbation. It has the effect of incorporating organic material in the sediments in finely divided form, making it available to the meio- and microbenthos.

The small amount of information available about the meiobenthos indicates that it is normally dominated by nematode worms and harpacticoid copepods. These have a variety of mechanisms for browsing on organic fragments, unicellular algae and bacteria, many of which are scraped from the surfaces of sediment particles.

The only renewable source of oxygen for the support of the meio- and microbenthos is from the water above the sediment. In fine-grained sediments the movement of water within the sediment is restricted, and a few centimetres below the surface one encounters anaerobic conditions, usually marked by the presence of black hydrogen sulphide. The meiofauna is mostly confined to the upper aerobic layer of sediment, but in the lower zone there are anaerobic bacteria of many kinds, prominent among them being those which obtain their oxygen by the reduction of sulphate and use buried organic matter as a carbon source.

In addition to this diverse collection of organisms utilizing dead organic matter, there is a selection of animals which live by predation on others in the community. Large animals such as fish, crabs and lobsters roam freely over the bottom while starfish move slowly and often kill their prey by holding them in their tube feet while they evert their stomachs over them. Some of the gastropods bore holes in the shells of other molluscs, then insert the radula, while carnivorous polychaete worms use the cuticular teeth on an eversible proboscis. Since it takes many calories of prey to support one calorie of predator, it is inevitable that predators are less abundant than their prey.

6.3.2   BIOMASS OF THE BENTHOS

There is a rough proportionality between the food supply available and the biomass of organisms which is supported by it. Hence, the rules governing the food supply also govern the benthic biomass: it is roughly proportional to the primary productivity of the overlying water, but inversely proportional to the depth of the water column and the depth of the mixed layer. In a plot of benthic biomass against mean depth for 116 lakes in Europe and North America, Deevey (1941) showed clearly that very deep lakes never have a very large benthic biomass, but that shallow lakes cover the whole range from low to high biomass (Fig. 6.2). We must remember that both factors are operating simultaneously here: very deep lakes seldom have a high primary production in the first place, and the production has a high probability of being consumed before it reaches the bottom. The total range of Deevey's values is to a maximum of 1800 g m^{-2} (fresh weight) but an average value for the benthos of shallow lakes would be 100–200 g m^{-2}. For deep lakes it would be about 20 g m^{-2}.

The best data set for the benthos of the world's oceans has been obtained by the Russians, who have estimated that the mean biomass for the depth range 0–200 m (the continental shelves) is about 200 g m^{-2}. For 200–3000 m it is about 20 g m^{-2} and for depths greater than 3000 m it is only 0.2 g m^{-2}.

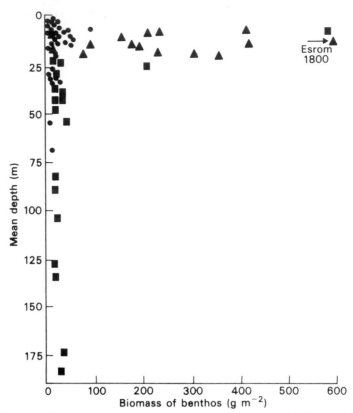

Fig. 6.2. Biomass of bottom fauna in relation to mean depth of lakes. ■ Alpine. ▲ North Germany. ● North America. [After Deevey 1941]

The benthos of rivers tends to consist of large numbers of rather small organisms. A river in Japan is reported to have contained 128 g m^{-2} dominated by caddis-fly larvae, and this is higher than in most rivers. Benthic faunas tend to depend heavily on the food supply washed into the river from the river basin, and there is a general trend towards increasing food supply as the water travels from the upland source towards the mouth. Hence there is also a tendency for the biomass of benthos to increase from source to mouth (Table 6.2). Occasionally, a dense population of molluscs can drastically change the picture. The River Thames when studied by the author, contained 121 g m^{-2} (fresh weight,

Table 6.2. Change in biomass of benthos in a Norwegian river. [After Økland 1963]

Approximate altitude	Distance from sea	Mean biomass of benthic invertebrates
(m)	(km)	(g)
50	13.5	3.0
20	8	5.6
10	5	21.5*
2	5	15.9

*Slight domestic pollution occurred above the 5-km station.

108

excluding shells) of freshwater mussels. Other species only amounted to about 10–15 g m^{-2} in total.

6.3.3 DIVERSITY AND STABILITY OF BENTHIC COMMUNITIES

If we are to understand processes in the benthos, we need to know whether benthic faunas are reasonably constant in their biomass and productivity from year to year. We have very little factual information on this point, but are able to reach some tentative conclusions on the basis of analogy with other better-known situations. Tropical rain forests and coral reef communities appear to persist for long periods with very little change. Both systems are notable for their high species diversity, and for a long time it was thought that their high species diversity was the cause of their constancy. It was argued that in a highly diverse community each predator has a wide choice of prey species, and each species is regulated by several predators. Hence, if one species is for some reason eliminated, neither its predators nor its prey will be too much affected. The system will be able to absorb the change with minimum disruption. However, attempts to explore mathematically the properties of complex ecosystems have thrown doubt on the validity of the relationship between species diversity and system stability, so the question is still open.

A further contribution to the debate is the suggestion that in complex ecosystems like tropical rain forest or a coral reef, there is a physical complexity of structure which provides a wide range of habitats for a diverse assemblage of organisms. In other words, the species diversity is a reflection of the massiveness and heterogencity of the structures, and has nothing particular to do with system stability.

Sanders (1969) provided a different point of view when discussing his monumental studies of marine benthos. He measured species diversity in his samples by what he called the rarefaction method (Fig. 6.3) in which he was able to plot the relationship of diversity to sample size. The method has some limitations, but the results seem to be clear. For polychaetes and bivalve molluscs, which dominated his samples from soft sediments, he found that species diversity was high in samples from the deep sea, below 300 m, but was low in samples from shallow water in temperate latitudes. Diversity was high in samples from shallow tropical habitats. He pointed out that high species diversity had also been found at great depth in major lakes such as Lake Baikal.

On the basis of this information Sanders (1968) suggested that the underlying cause of high diversity was the persistence of stable environmental conditions over long periods of time. In the depths of the ocean and of major lakes the annual temperature range is slight, and a small but constant input of organic matter has persisted for hundreds or even thousands of years. In shallow tropical waters stable conditions have persisted for a similar length of time. By contrast, shallow waters in temperate latitudes have a wide annual range of temperature, and are subjected to irregular and unpredictable changes in mean temperature and productivity.

According to Sanders' 'stability-time hypothesis' (Fig. 6.4) the communities in constant conditions have become 'biologically accommodated'. Biological stresses between species, such as intense competition, or non-equilibrium con-

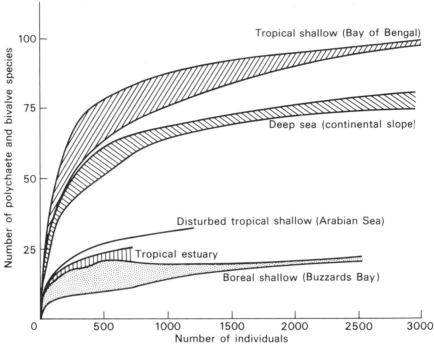

Fig. 6.3. Relationship between sample size and number of polychaete and bivalve species in various habitats. [From Sanders 1968]

ditions in predator/prey relations have been progressively reduced; more and more species have been added and have been accommodated, to give a complex, stable community which can be expected to persist so long as the constant environmental conditions persist. However, these organisms are now tolerant of a rather narrow range of conditions, and the community would not be dynamically stable under changing conditions. At the other end of the spectrum are the low-diversity communities of stressful, unpredictably fluctuating environments. Environmental stress reduces the community to a smaller number of

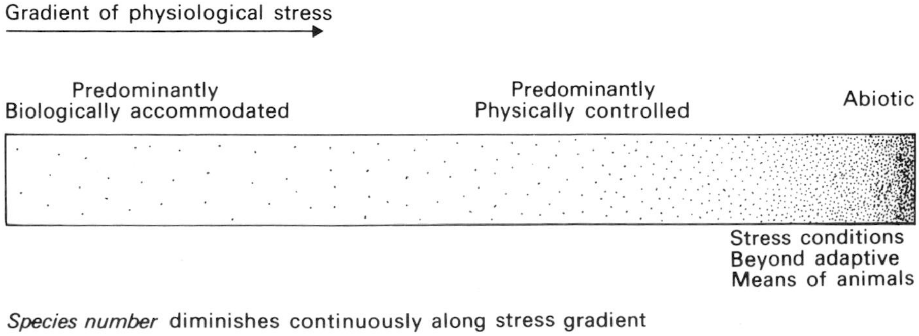

Gradient of physiological stress

| Predominantly
Biologically accommodated | Predominantly
Physically controlled | Abiotic |

Stress conditions
Beyond adaptive
Means of animals

Species number diminishes continuously along stress gradient

Fig. 6.4. Diagram illustrating the stability-time hypothesis of species diversity. [From Sanders 1968]

110

species able to tolerate a wide range of conditions, and able to reproduce rapidly when conditions are favourable.

While the details of the mechanisms giving rise to high and low diversity are still a matter of debate it is now clear that benthic communities at great depth experience relatively constant conditions and under these circumstances develop a high species diversity and are probably rather constant in composition over long periods. Shallow-water communities by comparison, have a lower species diversity and are likely to show marked fluctuations in biomass and community structure from year to year. The stability-time hypothesis predicts that since the annual temperature ranges of shallow waters on the western sides of continents are smaller than on the eastern sides, the western communities will have a greater diversity. This has been found to be true.

6.3.4 SOME CONSEQUENCES OF ENVIRONMENTAL STRESS

As stated in Chapter 1, temperature stratification of water bodies is a common occurrence, and may lead to greatly reduced oxygen concentrations in the lower layers. When this occurs, species unable to tolerate low oxygen conditions for a given length of time are eliminated, and benthic diversity is correspondingly reduced. Tubicid worms are among the most tolerant of low oxygen conditions and may be found in temporarily anaerobic environments in both fresh water and the sea. The larvae of chironomid midges are also tolerant of anaerobic conditions. However, low species diversity does not necessarily mean low population density. The benthos of highly productive, relatively shallow lakes may contain tens of thousands of tubificids ands and chironomids per square metre (see p. 115).

In running water, an analogous condition may be produced by the discharge of raw or partially treated sewage into a stream bed. The oxygen requirements of the bacterial populations which develop as it moves downstream may produce very low oxygen concentrations in the river, and greatly reduce species diversity (Fig. 6.5). Similar reductions in species diversity are found in lake and coastal benthic communities subjected to pollution stress, and in estuaries where animals are subject to salinity stress.

6.4   Productivity of benthic communities

Since production of new tissue and its transfer to other subsystems is one of the key ecosystem functions of the benthos, we should now consider the relationship between biomass and productivity. About 40 years ago it was shown by K.R. Allen that the food requirements of the trout population in a New Zealand stream were many times greater than the amount of benthos present at any one time. He concluded that the benthic community must replace its biomass several times in a year. Before that time, few people had thought much about the dynamics of secondary production, or of the relationship between mean biomass and annual production. The biomass that we see at any one time is in constant change: new material is being added by the growth of individuals and by reproduction, while existing biomass is being removed by predation, or by death and decay.

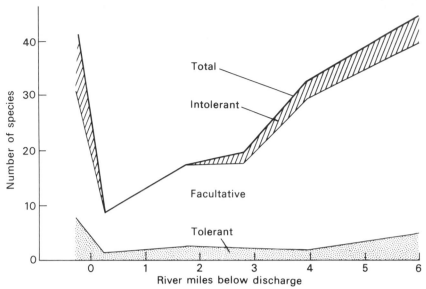

Fig. 6.5. Number of macrobenthic species (classified as tolerant, facultative or intolerant) at various points in the Red Cedar River, Michigan. Industrial wastes enter at mile 0. [After Willson 1967]

To get a clear picture of the dynamics of the situation, it is best to start with an organism that has a well-defined breeding period, and follow the fate of a cohort of newly-hatched young. Fig. 6.6a from Peer (1970) shows the results of sampling a cohort of the marine worm *Pectinaria hypoborea* on five successive occasions. As time progressed, the numbers decreased by predation while the mean weight of the animals increased as they grew. Curves of this kind were first produced by K.R. Allen and have come to be known as Allen curves. From this information, three parameters of the population can be calculated for each interval between samples: the production (P) from the average number present and their average weight increment; elimination (E) from the number lost from the population and their average weight; and average biomass present (B) from the product of average numbers present and mean weight of an individual. The change in biomass (ΔB) between sampling dates is given by the difference between material added by growth production and material lost by predation, i.e.

$$\Delta B = P - E \qquad \qquad (Equation\ 6.i)$$

Conversely, production is equal to the amount removed by predators, plus any biomass change, i.e.

$$P = E + \Delta B \qquad \qquad (Equation\ 6.ii)$$

Fig. 6.6b shows how these parameters changed with time in the *Pectinaria* population.

In situations where breeding and recruitment of young occurs over a protracted period, it is often not possible to identify cohorts and estimate growth

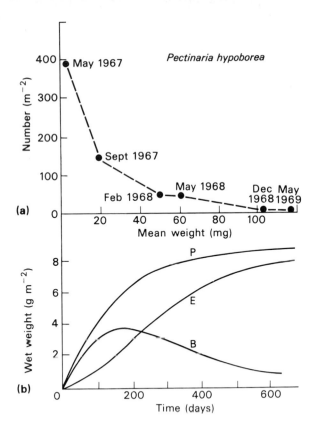

Fig. 6.6. (a) Plot of numbers against mean weight (Allen Curve) for a cohort of a marine worm; (b) changes with time of cumulative production (P), cumulative elimination loss (E) and biomass (B). [From Peer 1970]

and mortality. In those situations the only course open is to determine experimentally the growth rate of organisms of various sizes, and estimate production by multiplying numbers present by rate of growth. It is easy to see that calculation of productivity for all the species of a population is a tedious undertaking. In the search for some method of indirect calculation, people have tried to see how the ratio $P:B$, which is an expression of the number of times the biomass turns over in a given time interval, varies according to species and environment. Waters (1969) made a theoretical analysis of Allen curves with differing patterns of mortality and growth. He showed that within *the life span of a cohort* the P/B ratio normally lies in the range 3 to 6. Let us take 5 as a representative value. It follows that if the life span of an animal is 1 year, annual production is about five times the biomass present, i.e. $P:\bar{B} = 5$ on an annual basis. For organisms having 2 generations per year, $P:\bar{B} = 10$, and for those living 2 years $P:\bar{B} = 2.5$. This hypothesis was tested by the author by measuring the $P:\bar{B}$ ratio of numerous invertebrates in the River Thames, and found to be approximately true. Zaika examined the relationship for 22 species of aquatic mollusc and he too found it to be approximately true, and those marine benthic invertebrates for which the calculations have been made also fit the pattern. Obviously there are variations due to temperature (Fig. 6.7) food supply, crowding (which may be a function of predation) and so on, but the general outline of our understanding of $P:\bar{B}$ ratios is already in place, only awaiting refinements.

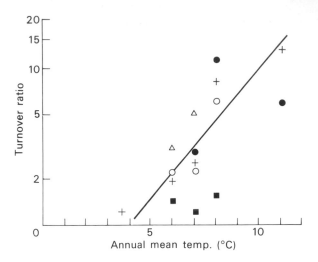

Fig. 6.7. Turnover ratio (P/B) of the macroinvertebrate community and of various taxonomic groups in Lake Ontario plotted against annual mean temperature. Line fitted by eye.
△ Chironomids.
● Oligochaetes.
■ Sphaeriids.
○ Crustaceans.
+ All groups.
[After Johnson & Brinkhurst 1971]

6.4.1   PRODUCTION AS ENERGY FLOW

When organic matter enters a benthic community it may be consumed by macrofauna, meiofauna or microbes. There are complex and poorly understood food webs within the benthos but the final outcome is that much of the energy content of the food is respired and lost as heat, while a proportion is stored in the bodies of animals as secondary production, to be removed from the system by predators. A good way of studying such processes is to express them in energy terms. We have seen that the input to the benthos below the photic zone varies according to the primary productivity and depth from very low levels up to about 200 g C m^{-2} yr^{-1}, which is equivalent to about 2000 kcal m^{-2} yr^{-1} (or 8374 kjoules m^{-2} yr^{-1}). In shallow euphotic situations primary production may be several times greater than this, but much of it is carried by wave action or currents away from the area in which it is produced.

If 200 g m^{-2} (see p. 107) is a representative biomass figure for the continental shelves, we may take this as very roughly equivalent to 160 kcal ($=670$ kjoules) m^{-2}. Considering that molluscs and polychaete worms with life histories of 2 or more years predominate, a P/B ratio of 2 might be appropriate for a first approximation. This gives a benthic productivity of 320 kcal ($=1340$ kjoules) m^{-2} yr^{-1}. In running water, benthic secondary production estimates range from 70 to 614 kcal m^{-2} yr^{-1} (Mann 1975) and thus appear to be of the same order of magnitude as the marine estimates. There are very few estimates of the productivity of a whole benthic community in a lake. We saw earlier that benthic biomass in lakes of different types lies in the range 20 to 200 g m^{-2} fresh weight. Applying a P/R ratio of 5 and a conversion of 1 g $= 0.8$ kcal, we obtain estimates of productivity of 80 to 800 kcal m^{-2} yr^{-1} which is again similar to figures from other habitats.

Hence, from an input of up to 2000 kcal m^{-2} yr^{-1} we can see evidence of benthic production of the order of 300 to 400 kcal m^{-2} yr^{-1}, much of which is made available to predators such as fish which remove it from the benthic subsystem. The remainder of the energy is released as heat when the organic matter is decomposed during its passage through the intricate food web com-

114

prising microbes, meiofauna and macrofauna. In the course of this process, the organic compounds are mineralized, except for a residue, usually small, which escapes decay and is buried. It may eventually be fossilized.

6.4.2 EFFECT OF STRATIFICATION ON PRODUCTIVITY

In lakes which have periods of stratification accompanied by deoxygenation of the deep waters, the diversity of the fauna is reduced but there may be a great abundance of organisms adapted to surviving a limited period of low oxygen concentration. For example, in Lake Esrom in Denmark it was shown that the midge larva *Chironomus anthracinus* was present in densities up to 40 000 m^{-2}. Growth was very slow during anaerobic periods, and tended to accelerate markedly when oxygen returned to deep waters a the autumn mixing (Fig. 6.8).

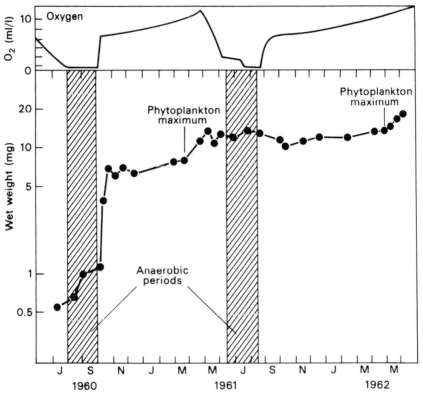

Figure 6.8. Growth of 3rd instar larvae of *Chironomus anthracinus* at a depth of 20 m in Lake Esrom, Denmark. [After Jónasson & Kristiansen 1967]

Other factors influencing growth were temperature and the availability of food. Very little growth occurred during the winter months, but there was burst of growth when phytoplankton from the spring bloom began to sink to the bottom. These larvae grew at about 10 mg per year, so for every 1000 animals there is a production of about 10 g (about 8 kcal). Hence, even under these relatively unfavourable circumstances, benthic production is not negligible. In the same habitat there also lived a dense population of the tubificid worm *Potamothrix hammonensis*, which contributed a further $10 \, \mathrm{g \, m^{-2} \, yr^{-1}}$ fresh weight of secondary production. The combined effects of poor oxygen conditions and competition with *Chironomus* were to slow down growth and delay breeding until the third or fourth year of life. Under these circumstances the P/B ratio was only 0.8. Anaerobic conditions may also occur in sheltered marine habitats but there is no comparable study of benthic productivity.

6.5   Regeneration of nutrients from the benthos

As was mentioned earlier, the mineralization of organic matter sinking to the bottom, and the return of dissolved ammonia, nitrate, phosphate, etc. to the water column is a key ecosystem role of the benthos. In situations where adequate dissolved oxygen is supplied to the bottom at all times, most of the mineralization occurs at the sediment surface and in the aerobic layer of the first few centimetres of sediment. As dead plant material, dead animals and faecal pellets sink to the bottom they are rapidly colonized by bacteria and fungi. The organic matter that is decomposed directly by microorganisms eventually becomes carbon dioxide and nutrient salts. That which is ingested and digested by meio- and macrofauna is transformed to excretory fluids which are rich in ammonia and phosphates. Finally, as we have seen, faecal pellets become well mixed with the surface sediments, may be reingested several times by invertebrates, and by this means even the refractory organic substances are eventually mineralized. That fraction which is carried down in the anaerobic sediments is acted upon by a variety of organisms capable of anaerobic fermentation or of sulphate reduction.

In a study of nutrient regeneration from the sea floor in Naragansett Bay, Rhode Island, U.S.A. it was shown that the flux of ammonia varied from very low levels in winter to a maximum of about $400 \, \mu\mathrm{M \, m^{-2} \, hr^{-1}}$ in summer. The strong temperature dependence is not surprising since the process is brought about by the metabolism of microbes and invertebrates. Up to this time, it had not been known whether the nutrients used by phytoplankton in coastal waters were mainly regenerated from the benthos, or whether currents brought in a fresh supply from deep water. The results of this study in Narragansett Bay were fed in to a computer model of the hydrology of the bay, and the flux of ammonia was found to be sufficient to account for uptake by the phytoplankton. The same study gave indirect evidence of a strong flux of organic nitrogen, implying that some organic material in solution diffused back into the water column before it had been fully degraded by bacteria.

The turnover time for dissolved nitrogen in the interstitial waters of sediments appears to be quite short; there appears to be rather little present at any one time, relative to the flux. The reverse seems to be true for phosphorus. Quite large amounts are present in most sediments, and in well-aerated environments there appears to be a slow continuous exchange with the overlying water. In shallow marine habitats such as salt marshes and seagrass beds, marine angiosperms draw freely on the phosphate reserves of the sediments and then liberate them in the water column when their tissues die and decay. Some people refer to such plants as 'phosphorus pumps'.

116

The seasonal pattern of events is quite different in environments where the deep water becomes anaerobic for part of the year. Low oxygen conditions are accompanied by a rapid release of dissolved ammonia, phosphate, and ferrous ions. During the aerobic phase the iron is present in the sediments in the ferric state, and has the power of adsorbing large numbers of ammonium and phosphate ions. It also forms a layer which blocks the diffusion of nutrients from deeper parts of the sediment. Onset of the anaerobic phase reduces the ferric ion to the ferrous state, liberating the adsorbed nutrients and freeing others from the deeper sediments.

6.6   Life histories of benthic animals

The majority of benthic animals are relatively immobile. Many, such as coelenterates, sponges, and barnacles live firmly attached to a solid substrate, filtering food from the passing water. Others, such as flatworms, leeches, molluscs and various kinds of arthropods move slowly over a limited area in search of food. The annelids, bivalve molluscs and the chironomids normally burrow into the sediments and remain in one place for long periods. The big disadvantage of these modes of life are that they restrict the species in such activities as colonizing new habitats, or spreading favourable gene mutations and combinations. To overcome this disadvantage, a high proportion of benthic animals evolved planktonic larval stages. In the sea, about 70% of all species have planktonic larvae; the proportion rises to about 80% on the continental shelves of temperate and tropical waters, but it is lower in the arctic and in the deep ocean.

Liberation of eggs or larvae into the plankton involves the risk of their being carried to areas unsuitable for adult life, and that while in the plankton they may be subject to heavy mortality from predators. One strategy for maximizing the chances of success for the species is to produce very large numbers of larvae. Since the parent has limited resources, it is inevitable that such larvae will each have a very small food reserve, and will have to feed themselves from early in development. Hence, success also depends on the larvae being liberated at a time when food is plentiful, or being carried by currents to areas where food is plentiful. In temperate climates most larvae are liberated at the time of the spring bloom of phytoplankton.

An alternative strategy is to give each larva a generous food reserve that will carry it through to a later stage of development. When this is done, the number of larvae produced is necessarily much smaller. This strategy is favoured in habitats such as the arctic and the deep ocean, where food resources may be severely limiting.

Species producing large numbers of larvae are characterized by spectacular variations in the numbers of young recruited to the population in different years. If things go well for the larvae, and ocean currents bring them to a favourable habitat at the time when they are ready to settle on the bottom, there may be a massive recruitment of young. In other years there may be almost total failure of recruitment in that same area. The result is a wide fluctuation in population density. Many bivalve molluscs fall in to this category. Recruitment to populations of animals having eggs with good food reserves (yolk) tends to be much more constant from year to year. In the coastal marine environment crabs and lobsters, in which the yolky eggs are brooded by the females, would be good examples of the second category. The recruitment to commercial stocks of these crustaceans seems to be fairly constant from year to year, whereas recruitment to stocks of scallops is highly irregular.

Opinions differ about how far planktonic larvae normally travel and to what extent they bring about genetic mixing of stocks. One author maintains that 90% of larval populations remain in the general area of the parent stock, while another claims that there is a high probability that planktonic larvae of some molluscs, coelenterates, annelids and crustaceans may survive long enough in ocean currents to cross from one side of the Atlantic to the other.

In rivers, where there is unidirectional flow of water towards the sea, planktonic larval stages are less favoured. Aquatic insects, leeches and gastropods tend to cement their eggs to a solid object; amphipod and isopod crustaceans brood their eggs under the body of the female, mussels retain the eggs in the mantle cavity, and so on. This habit tends to persist even in lake-dwelling forms, for there is no sharp division between lakes which are little more than a local widening of a river, and lakes which have very little inflow and outflow relative to the volume of the lake.

There is much we still do not understand about the distribution of benthic invertebrates. We do not know how invertebrates colonize isolated bodies of water. It is suggested that small animals, or the young stages of larger animals, may travel among the feathers of aquatic birds, or in mud on the birds' feet. Whatever the mechanism, it is fairly efficient, for when some water bodies were formed *de novo* in England, they all had representatives of the major aquatic groups within 8 years.

7

The Nekton:
Production and Migration Patterns

F.R. HARDEN JONES

7.1　　Introduction

This account describes the movements of the nekton in relation to production cycles, and it begins with a reminder of some terms. Benthic organisms live in or on the bottom while pelagic* organisms live freely in the water column (Chapter 1). Here the plankton may be contrasted with the nekton, the former drifting passively whereas the latter appear to choose their own path or maintain a particular distribution either on account of their swimming ability or of some

Table 7.1. Commercial fishery landings in 1977. Estimates based on F.A.O. (1978). The landings of marine mammals have been estimated from numbers of whales and seals reported in the F.A.O. statistics and the weight data given in various papers (e.g. Lockyer 1976).

Landing group	Landings	
	In thousands of metric tonnes	As % of World Total landings
Seaweeds	1 490	2.03
Crustacea		
inland	45	0.06
marine (less krill)	2 162	2.94
krill	123	0.17
Fish		
inland	9 769	13.30
diadromous (salmons, etc)	1 554	2.11
marine	53 963	73.42
Turtles	7	0.01
Bird guano	40	0.05
Marine mammals	517	0.70
Inland waters (total)	10 758	14.64
Marine (total)	62 743	85.36
World Total	73 501	100.00

*The pelagic state is often contrasted with the demersal, as for example, pelagic and demersal eggs, species, or fisheries. Thus demersal eggs, such as those of the herring, are laid on and stick to the bottom in contrast to the pelagic eggs of the cod which are freely liberated into the water. Pelagic species (krill, squid, herring, mackerel, tuna, blue whale) spend much or all of their time in midwater and are mainly or entirely independent of the bottom. Demersal species spend a significant part of their adult life near the bottom (shrimps, lobsters, cod, haddock), on the bottom (plaice, dabs) or even buried in it (sole). So in a demersal fishery the catch is taken close to or on the bottom, whereas in a pelagic fishery the catch is taken in midwater.

special feature of their behaviour. The nekton include, among the invertebrates, the larger crustaceans and molluscs; and among the vertebrates, fish, live-bearing sea snakes, and the cetaceans. To these could be added those vertebrates whose reliance on the sea is complete except for the needs of reproduction: the marine turtles and egg-laying sea snakes among the reptiles; the pinnepedes (seals) among the mammals; and a number of oceanic birds—including albatrosses, petrels, shearwaters, boobies, gannets, tropic and frigate-birds—whose independence of the land and dependence on the sea parallel those of their reptilian counterparts.

The extent to which the nekton and other groups contribute to the commercial fisheries is shown in Table 7.1. The marine catch is six times that from inland waters; marine plants are harvested only as seaweeds and then necessarily in shallow waters; 95% of the landings of crustaceans are demersal species (crabs, lobsters, prawns and shrimps) and the only representative of the pelagic species is the Antarctic krill *Euphausia superba**; squids account for almost 70% of the cephalopods landed; and fish dominate the landings of marine vertebrates, mammals now making up less than 1% of the total.† Chapter 8 considers the commercial catch of nektonic species in more detail.

7.2   Movements and water currents

The distinction between plankton and nekton is almost entirely one of size (Chapter 1); as adults the larger animals are strong enough to move independently of the currents whereas their young stages, or smaller animals, are often physically unable to do so. Some members of the plankton maintain their distributions by setting the drift of a surface current against that of a deeper counter-current, the animals migrating vertically between them on a seasonal or diurnal (24 hr) basis. On the continental shelf, where the tidal streams are stronger than the oceanic currents, inshore and offshore movements may be brought about by moving from the bottom into midwater on the flood or ebb tide, the vertical migration having a semi-diurnal (12 hr) rhythm.

There is a wide range to the distances moved by members of the nekton. Some species have a restricted distribution throughout their adult life; for example, fish may stay in a particular section of a stream or on one coral reef. Other species make seasonal inshore and offshore movements of up to a few kilometres: prawns, shrimps, and many species of littoral fish fall within this category. More extensive movements on the continental shelves are undertaken by several species; for example by penaeid prawns such as *Penaeus duarorum* in the Gulf of Mexico, the Japanese flying squid *Todarodes pacificus* in the Sea of Japan, and herring, cod, hake and plaice in temperate waters; the distances covered may amount to several hundred kilometres during a year. Finally, some animals make very much longer journeys over distances of several thousand

---

*A few metric tons of the copepod *Calanus finmarchius* are taken in Norway (Wiborg 1976) and are used as a source of astaxanthin to give the desired red colour to the flesh of cultured salmonids.
†In the 1930s the annual whale catches of 2 to 3 million metric tons corresponded to 15 to 20% by weight of the total landings.

kilometres: examples include diadromous species such as salmons and eels which move between freshwater and the high seas; and tuna and whales which make transoceanic migrations.

But size by itself does not necessarily bring independence from the currents. Porcupine fishes, trigger fishes, and sunfishes attain a fair size but depend on their paired fins for propulsion: they are poor swimmers and are carried almost passively with the currents. Furthermore, swimming ability needs to be matched by some element of directionality in the animals' movements, a requirement which raises problems of behaviour and sensory physiology.

Aquatic animals often collect close to the temperature and salinity gradients which occur at fronts between water masses. The distances across the fronts are small, a few hundred or a few thousand metres, the aggregation taking place within a zone which is perhaps no wider than 10^3 or 10^4 times the length of the animal. These distributions are usually associated with scalar quantities—for example, gradients of temperature or salinity—rather than vector quantities, and it is unlikely that the underlying behavioural mechanisms could provide the basis for the more extensive movements involved in migrations during which the animal will cover several hundred or thousands of kilometres, distances equivalent to 10^7 to 10^9 times its body length. Directionality usually implies the use of vector quantities and dependence on an external reference point. For a demersal fish this could be provided by a touch or view of the seabed to initiate a response to a water current; an occasional glimpse of the shoreline or nearby mountains might allow a Californian gray whale to keep to a course parallel to the coast; and a sight of the sun might be sufficient to sustain a light compass reaction, and allow a fish close to the surface to maintain a steady course.

In shelf waters, however, even powerful swimmers such as adult plaice, cod, and dogfish, may use the tidal streams for transport in a manner similar to some invertebrates and younger fish, a semi-diurnal vertical migration from the bottom into midwater allowing the fish to be carried by one stream rather than another, the transport system both saving energy and adding directionality to their movements.

It is clear that the currents must have considerable influence on the distribution and life histories of the nekton, both in their young and older stages. In the main ocean basins the surface circulation is wind driven and its effect extends to at least 40–50 m and so often involves much or all of the euphotic zone. The zonal wind patterns arise from the interaction between the advective movements of the air—due to latitudinal and seasonal differences in solar radiation—and the deflecting effects of the earth's rotation: von Arx (1962) and Munk (1955) give very readable accounts of these matters. In the northern hemisphere the northeast trade winds drive the equatorial currents to the west, while in more northern latitudes the prevailing westerlies drive the water to the east. In high polar latitudes the prevailing, if erratic, easterlies again drive the water west. Under the influence of the zonal wind patterns the oceanic circulation resolves into several closed gyrals or circles: the subtropical and subpolar gyrals and the polar current in the northern oceans; and the subtropical gyrals, circumpolar West Wind Drift and the East Wind Drift in the southern oceans. A simplified chart of the surface currents of the oceans is given in Fig. 7.1. Substantial counter-currents may lie below the surface currents and in the subtropical gyres

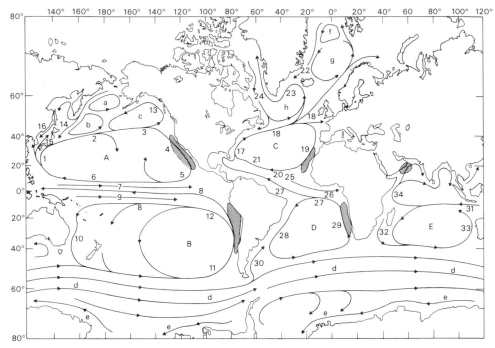

Fig. 7.1. A simplified chart of the surface currents of the oceans. Main subtropical gyrals: (A) North Pacific; (B) South Pacific; (C) North Atlantic; (D) South Atlantic; (E) South Indian. Other gyrals: (a) Bering Sea; (b) Western subarctic; (c) Alaskan; (d) Antarctic circumpolar; (e) East wind drift; (f) Greenland Sea; (g) Norwegian Sea; (h) Irminger Sea. Main ocean currents in the Pacific: (1) Kuroshio; (2) Kuroshio extension; (3) West wind drift; (4) California; (5) California extension; (6) North Equatorial; (7) North Equatorial counter; (8) South Equatorial; (9) South Equatorial counter; (10) East Australian; (11) Humboldt; (12) Peru; (13) Alaskan; (14) Oyashio; (15) Tsushima; (16) Liman; in the Atlantic; (17) Gulf Stream; (18) North Atlantic; (19) Canary; (20) North Equatorial; (21) Antilles; (22) East Greenland; (23) Irminger; (24) Labrador; (25) Equatorial counter; (26) Guinea; (27) South Equatorial; (28) Brazil; (29) Benguela; (30) Falkland. Main ocean currents in the Indian Ocean: (31) South Equatorial; (32) Agulhas; (33) W. Australian; (34) Somali. The circulation in the Indian Ocean is shown during the period of the S.W. monsoon (April to September). Important areas of upwelling in the eastern boundary currents (California, Peru, Canary, and Benguela) and off Arabia are indicated. [Partly after Bramwell 1977]

these are particularly important in the equatorial and coastal boundary areas.

While oceanic wind-driven currents predominate on the high seas, tidal influences are more important on the shelves. Fig. 7.2 identifies four distinct current regimes: oceanic currents off the shelf; the increasing influence of tidal currents on the shelf; river discharges adding to the ebb tide in estuaries; and in rivers the downstream flow (and see Fig. 10.1). The different regimes present different opportunities and problems to nektonic animals in terms of behaviour, movement and distribution. Migrants which have to contend with all four regimes (the anadromous freshwater-spawning salmons and the catadromous marine-spawning eels) might be expected to show marked changes in behaviour as they move from the high seas to freshwater and vice versa.

122

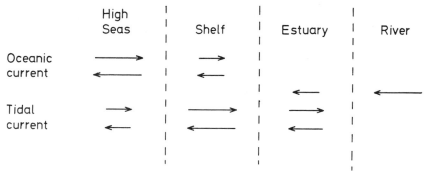

Figure 7.2. Current regimes in the aquatic environment. Oceanic currents are dominant on the high seas, tidal currents on the shelf. The length of the arrows indicates the relative strength of the currents.

7.3   Primary production and the nekton

Living resources in the sea ultimately depend on photosynthesis and the production cycles characteristic of arctic, temperate, and tropical latitudes have been described in Chapter 2. In the discontinuous cycles characteristic of high latitudes the peak rates of production lie within the range of 1 to 5 g C/m^2 per day and areas where such values are found include the shallow waters of the continental shelf, the temperate and subarctic waters of the Southern Ocean, North Atlantic, and North Pacific, and areas of divergence in the equatorial zones and along the polar fronts at the high latitude borders of the subtropical anticyclonic gyrals. The divergent action of the winds is particularly effective in the eastern boundary currents of the subtropical gyrals (the Californian and Canary currents in the northern hemisphere, the Peru and Benguela currents in the southern hemisphere) where deep, relatively cold, nutrient-rich water rises (or upwells) to the surface at speeds of 1 to 5 m a day to replace that deflected offshore under the influence of the prevailing Trade Winds (which blow parallel to the coast towards the equator) and the effect of the earth's rotation. The relatively deep (200 m) counter currents associated with the eastern boundary currents have already been mentioned: off California, the Deep Coastal Undercurrent extends from Lower California (25°N) to Oregon (40°N).

Upwelling is not continuous throughout the year and there is a shift in the centre of upwelling which moves polewards during the summer months, reflecting seasonal changes in the wind patterns. In the California current the centre of upwelling moves from Lower California (25°N) in March and April northwards to the Washington coast (45°N) in August and September; and in the Peru current there is a movement southwards from Cape Blanco (5°S) in May and June to the Ile de Chiloe (45°S) in November and December. Similar changes occur in the Canary and Benguela currents. Similarly the southwest monsoon (April to September) is associated with a seasonal upwelling along the coast bordering the Arabian Sea (see also pp. 35–6, 151, 185–6).

The rate of primary production in the upwelling areas is similar to the peak values found in temperate and arctic latitudes, and is five to ten times greater than that found in the open waters of the subtropical gyrals. The higher daily rates reflect the greater standing stock of phytoplankton which builds up during

123

the delay period between the onset of plant production and the time when the herbivorous zooplankton population has increased sufficiently to graze it down. In the higher latitude production cycles, the delay is reflected in the change in the standing stock of plants as the season progresses, whereas in the upwelling areas the standing stock increases and then decreases with distance from the coast as the surface waters move offshore and over deep water. The production cycles in upwelling areas differ from those in higher latitudes in one other respect which is likely to be significant: the upwelling cycles are probably three to four times longer than their counterparts in higher latitudes.

The geographical and seasonal variations in primary production should be reflected in the distribution and abundance of the herbivorous invertebrates, in particular the zooplankton (Chapter 3), and also in the distribution and abundance of the higher trophic levels—crustaceans, cephalopods, fish, mammals, and birds—some of which feed directly on the phytoplankton, or, more usually, on zooplankton, fish, and other animals which are themselves directly or indirectly dependent on plants. The charts included in the *Atlas of Living Resources of the Sea* (F.A.O. 1972) clearly show that there is, in general terms, such a relationship between primary production, zooplankton distributions, and the yields of the commercial fisheries. A quantitative estimate of the production to be expected at higher trophic levels can be calculated if the efficiency with which energy is transferred from one level to another is known. Thus the rate of production P at a particular trophic level can be estimated as:

$$P = BE^n$$

where B is the rate of primary production, E the ecological or transfer efficiency, and n the number of links or levels between the first trophic level (primary production) and that corresponding to P (Parsons 1976). Thus if the primary production was 5 g C/m^2 per day and 20% of the energy was passed from one trophic level to the next, the expected production at the second, third and fourth trophic levels ($n = 1$, 2, and 3) would be 1.0, 0.2, and 0.04 g C/m^2 per day respectively. If both the total area of production and the duration of the production cycle are known, the daily rates of production could be used to calculate the annual production. Ryther (1969) has done this for the world oceans and his results are summarized in Table 7.2, where the annual production of fish in the sea is estimated as 240 million tons, of which half could be produced in upwelling areas which occupy only 0.1% of the area of the oceans. While there may be some doubts about Ryther's choice of transfer coefficients (too high?) and his estimate of the number of links in the food chain in an upwelling area (too low?), such figures are useful in that they suggest a limit to the potential yield of fish at about 100 million tons a year (half the annual production) and they indicate where the bulk of the catch is likely to be taken.

The richness of the four upwelling systems associated with the eastern boundary currents—the California, Peru, Canary and Benguela—is well documented (Cushing 1971b) and continues to attract great interest (Boje & Tomczak 1978). The upwelling ecosystems show remarkable parallels in their faunas which include sardines, anchovies, mackerels, hakes, tuna, lantern fish and squid. The larval fish and smaller species feed on phytoplankton and zooplankton; the younger stages of larger species on zooplankton and smaller fish; and the

124

Table 7.2. Estimates of fish production (wet weight) for the World Oceans divided into three main provinces. [After Ryther 1969]

Marine province	Percentage of ocean area	Area (millions km²)	Primary production		Number of links in food chain	Transfer efficiency between trophic levels (%)	Estimated annual fish production, millions tons
			Annual rate (g C/m²)	Annual total millions tons carbon			
Open ocean	90.0	326.0	50	16 300	5	10	1.6
Coastal Zones	9.9	36.0	100	3 600	3	15	120.0
Upwelling areas	0.1	0.36	300	100	1½	30	120.0
Totals	100.0	362.36		20 000			241.6

older stages of the larger species on other fish. To this food web can be added the fish-eating birds (pelicans, boobies, and cormorants) and their vast deposits of guano after which many a white cape is named. While the F.A.O. catch statistics do not give the detail which allows the catches from the main upwelling areas to be identified, the catches from regions 77, 87, 34 and 47 (which include the California, Peru, Canary and Bengula upwellings respectively) amounted to 12.4 million tons in 1977, equivalent to 23% of the marine fish catch of 54 million tons. So the catch statistics tend to confirm the productivity of the areas deduced from measurements of primary production and food chain studies even in a year when the catch of the Peruvian anchoveta was only 800 000 metric tons, the lowest since 1958 and less than a sixteenth of the peak landings of 1970.

7.4 Production cycles, spawning seasons, and migration

Many larvae hatch with variable but limited reserves in their yolk sacs and if they are to survive they must find suitable plants and animals in densities on which they can feed effectively before their own reserves are depleted. Fish larvae usually start to feed on plant cells and go on to take infusoria, rotifers, and the small larvae of invertebrates. In tropical latitudes there is probably always enough food available to ensure the survival of moderate broods of larval fish. But as standing stocks are low, and production is continuous, spawning might be more successful if spread over several months.

In contrast the discontinuous production cycles in temperate and polar latitudes are associated with substantial but relatively short-lived standing stocks of phytoplankton and zooplankton which would sustain heavy and successful spawnings if they were so timed that the majority of the larvae were looking for food when it was most abundant. The productive upwelling areas in lower latitudes provide similar opportunities for large stocks of fish in tropical and subtropical waters. But while the production season is long, there is a seasonal movement towards the poles during the summer and here again reproductive success is most likely to favour those species whose eggs hatch in the right place at the right time.

As the young stages are often carried passively by the currents, their chance of survival would be increased if the eggs were spawned in a position from which the larvae would be carried to a favourable nursery area or ground. And the chance of survival will be greater for those fish whose subsequent behaviour enables them to eat more, grow faster, and avoid predators with greater success than other members of the brood.

Some of the nekton migrate: they come and go with seasons and migratory behaviour would appear to be one of several features in their life histories directed towards reproduction. The Grand Strategy appears to be that of ensuring a sufficient number of viable offspring to maintain the population up to the limit, in numbers or in weight, that can be fed. Migration allows the members of a particular stock to exploit, on a seasonal basis, the resources of different areas. More food could lead to greater survival, particularly among the larval and younger stages; a faster growth rate would allow more juveniles to escape predation because they would spend less time at risk in a vulnerable size range; the better-fed survivors might so increase their length at age as to mature

126

earlier than normal; and those mature individuals which enjoy an abundant food supply could produce more eggs up to a predetermined limit for a species of a particular size, or increase the reserves of a proportion or all the eggs released.

Increased survival among the larval and younger stages, earlier maturity, and an increased fecundity will all contribute to an increase in the population. If numbers are so dependent on the availability of food, migration could be regarded as an adaptation towards abundance and both the migration patterns and the life histories of the fish should be closely related to the regional production cycles.

In temperate waters the movements of migratory fish can be reduced to a simple triangular pattern in which the movements of the young and adults are linked together (Fig. 7.3). In general terms these movements are related to the

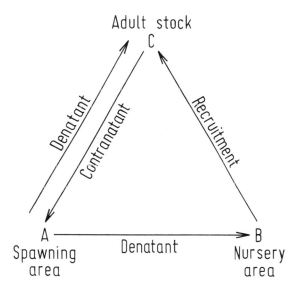

Figure 7.3. The migratory movements of young and adult fish linked together in a triangular pattern.

water currents and the young stages usually drift passively from the spawning area to the nursery area where they grow up. The spawning migration of the adult fish has been described as an active movement against a prevailing residual current and the return of the spent fish as a passive movement with the current. These two movements, the first active and *contranatant*, the second passive and *denatant*, are part of the traditional theory of fish migration: the adult fish feed in one area, winter in a second, and spawn in a third. Thus the migratory movements of the adults can also be represented by a triangular pattern, the sequence of seasonal movements around the migratory circuit being from feeding area to winter area to spawning area for spring spawners, and feeding area to spawning area to winter area for autumn spawners.

When on migration fish follow well-defined routes at regular intervals and the feeding, wintering and spawning areas occupied by a stock are stable over a period of at least several decades. Thus the fish return to places formerly occupied: in other words, they home. Consequently the populations of some species are grouped into separate stocks. There are, for example, stocks of cod at Georges Bank, Newfoundland, Labrador, West and East Greenland, Iceland, Faroe Bank, Faroe, the Barents Sea, the Norway coast, the North Sea, the Baltic, the English Channel, and the Irish Sea. While there may be at times significant mixing between the cod at Greenland and those at Iceland, and there may be more than one stock of cod in the North Sea, the different stocks are more or less independent units between which there is little mixing and in some cases at least, demonstrable genetic differences.

The distribution of a particular stock, i.e. the geographical extent of the migration circuit, can often be identified with a well-defined regional oceanic or tidal current system and individuals which stray beyond its boundaries are lost to their parent stock so far as reproduction is concerned.

But the number of species that make extensive migratory movements is relatively few. For example, there are probably no more than two or three hundred species of migratory fish whereas there are about 19 000 known or recognized fish species of which 7000 are found in freshwater. Those species which do migrate, however, are often of great importance to commercial fisheries and thereby to man (see p. 138). Fish are not spread uniformly over the world and there are fewer species at higher latitudes. Thus the number of polar species is limited to a few hundred; there are many more species in temperate waters; and the greatest number—perhaps over 70% of the total— is found in tropical waters where over 500 species may be found on a single coral reef. Other members of the nekton show a similar relation between latitude and the number of species.

7.5 Migratory movements of the nekton

Migratory movements representative of the nekton may now be considered against the background of the water movements outlined above and the production cycles described in Chapter 2. The examples chosen are of species of known or potential commercial importance.

7.5.1 THE ANTARCTIC KRILL

Euphausia superba (Eucarida, Crustacea) is the largest of the Antarctic macroplankton, and attains a length of 5 to 6 cm at the end of its second year of life. It is included with the nekton because the adults can swim at 15 cm s^{-1} for at least several hours.

Everson (1977) reviews the main features of the biology of the krill and its commercial potential. The krill dominates the ecology of the Southern Ocean. It feeds mainly on phytoplankton and in turn is eaten by squid, fish, seals, whales, and birds; the crabeater seal (*Lobodon carcinocephalus*) and baleen whales (blue, fin, sei, humpback and minke) depend upon it.

The krill has a circumpolar distribution within the Antarctic convergence (Fig. 7.4). In the West Wind Drift there are major concentrations in the Scotia Sea and the Weddell Drift; and in the East Wind Drift from Dronning Maud Land (0°) to Queen Mary Land (120°E), north of the Ross Sea (180°) and in the Bellingshausen Sea (90°W). It is not clear if these concentrations are self-

128

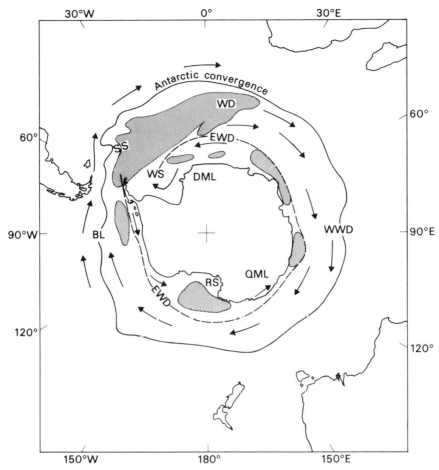

Figure 7.4. The distribution of the Antarctic krill within the Antarctic convergence. Abbreviations: *Sea areas:* SS (Scotia Sea); WS (Weddell Sea); WD (Weddell Drift); WWD (West Wind Drift); EWD (East Wind Drift); RS (Ross Sea); BS (Bellingshausen Sea); *Land areas:* DML (Dronning Maud Land); QML (Queen Mary Land). The northward boundary of the East Wind Drift is indicated by a broken line. [After Everson 1977]

maintaining stocks and the mechanism by which they hold their distribution against the set of the prevailing currents is not known: as yet there is no evidence to show that the krill compensate for the northeasterly movement of the Weddell Drift by making a seasonal movement to deeper water during the Antarctic winter and so returning to the south in the Warm Deep Layer.

Krill spawn during the Antarctic summer in January–March when they are 2 years old. Females release 2 to 4 thousand eggs, diameter 0.7 mm, which at first sink to 500–1000 m depth before rising towards the surface where development is completed. Adult krill live in the top 200 m, at least during the summer. They are extremely abundant and occur in dense swarms estimated to weigh from 10 to 50 metric tons. In the Antarctic summer these swarms are exploited by baleen whales. As the number of whales has been much reduced in the last 20 years, attention has been given to the possibility that there may be a substantial annual surplus of krill (up to a 100 million tons) some of which has undoubtedly been taken up by the crabeater seal whose numbers appear to have increased dramatically in the last 20 years. It has been suggested that the Southern Ocean might sustain a substantial pelagic fishery for krill (see Chapter 8), but its development would depend on the cost of maintaining a large fleet of midwater trawlers in the Antarctic for a 3- or 4-month season, the catch rates, and market potential. There could be strong arguments for harvesting the krill at a higher trophic level as in a well-managed whale fishery, or in an entirely new one. For example, could salmon—which are not indigenous to the Southern Ocean—be introduced to Tierra del Fuego? If they became established in the West Wind Drift and a proportion of the adults returned to the release area, the cost of harvesting would be much reduced. This superficially attractive proposal could run into difficulties if fish-eating seals (Elephant and Weddell) acquired a taste for salmon.

7.5.2 CEPHALOPODS

The pelagic cephalopods are voracious feeders and squid eat 15 to 20% of their body weight a day. Their stomach contents include animals from three trophic levels (primary consumers: copepods, chaetognaths and small euphausids; secondary consumers: amphipods, shrimps, large euphausids; tertiary consumers: cod, haddock and other squid) and they feed selectively from a wide range of organisms smaller than themselves. Squid are very abundant—a conservative estimate for a potential world catch is 10 million tons—and it is therefore somewhat surprising to find very little is known of their biology, and such information as is available is virtually restricted to the ommastrephid squids *Illex illecebrosus* (in the northwest Atlantic) and *Todarodes pacificus*; *Todarodes* is the common Japanese flying squid of which 300 to 700 thousand tons are caught each year in a partly mechanized jig fishery. *Todarodes* grows very fast, lives for a year, then spawns and dies. In Japanese waters (Kasahara 1978) this squid spawns in summer, autumn and winter in the northern part of the East China Sea (32°N, 127°E). Each female lays 300–400 thousand small eggs, probably on the bottom in semi-buoyant clusters. The eggs hatch in 5–10 days and the larvae are carried to the northeast into the Sea of Japan by the warm Tsushima current and along the east coast of Kyushu, and Honshu by the warm Kuroshio current (Fig. 7.5). The young squid grow rapidly and the best catches are taken along the lines of the fronts between the warm north-going and cold south-going currents (Tsushima and Liman in the Sea of Japan, the Kuroshio and Oyashio to the east of Honshu). At present the winter spawning group is the most important to the Japanese fishery and in September these squid reach the northern limit of their distribution to the west of Sakhalin (as far as 51°N) and close to the Kurile Islands (45°N). The return migration commences in October and copulation occurs on the way to the spawning grounds to the southwest of Kyushu. At the end of their life cycle the females reach a mantle length of 25 cm and a weight of about 300 g.

While many details of the migration are not known, the distribution of the squid and the pattern of movement recall that shown by pelagic fish, such as tuna and herring, in relation to frontal systems and oceanic currents. In its annual migratory circuit the Japanese squid covers a distance of about 4000 km.

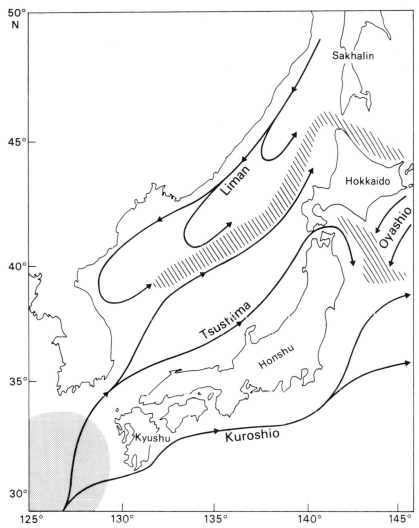

Fig. 7.5. The water currents in the Sea of Japan in relation to the spawning area (stippled) and movements of the common squid *Todarodes pacificus*. The frontal areas between the warm and cold water currents are cross-hatched.

7.5.3 SALMON

Salmon (genus *Oncorhynchus* in the Pacific and genus *Salmo* in the Atlantic) spend most of their lives in the sea and migrate to freshwater to breed; such a life history is termed anadromous. The sockeye or red salmon, *O. nerka*, lays a relatively small number of large yolky demersal eggs in Alaskan and British Columbian streams. Mature sockeye of the Fraser River, British Columbia, return to streams tributary to lakes in the autumn, spawn on gravel beds in December and die in freshwater. The eggs hatch in early spring. The young fish stay in their freshwater nursery for a year or more before making a spring migration downstream to the sea as smolts. After two winters at sea the mature adults, usually in their fourth year of life, return to freshwater to spawn.

131

Tagging experiments have shown that salmon that return to the Fraser and other rivers south of the Aleutian Islands are contained within the Alaskan gyral. While only 10% of the original downstream migrants return, 90% or more of those that do so are believed to spawn in their parent stream.

7.5.4 EELS

The American and European eels (*Anguilla rostrata* and *A. anguilla*) of the North Atlantic are catadromous: they spend most of their lives in freshwater and migrate to the sea to breed. The two species are very similar in appearance but individuals can be separated by their mean vertebral counts of 107 ± 1 and 115 ± 1 respectively. Both species are believed to spawn from January to May in the Sargasso Sea. They produce many small pelagic eggs which develop into leaf-like larvae (the leptocephali) which drift for $1\frac{1}{2}$ to $2\frac{1}{2}$ years in the Atlantic current towards the continental shelves. Here they metamorphose into elvers which swim up the rivers and so into freshwater where as yellow eels they feed and grow. The feeding areas of both species cover a wide range of latitude (20° to 64° in the New World, 22° to 70°N in the Old) and eels in northern latitudes grow more slowly than those in southern latitudes. As the onset of sexual maturity depends mainly on size, the mean period of freshwater residence of the downstream autumn migrants ranges from 8 years in the south to 12 or more years in the north. After a metamorphosis to silver eels—which have some features characteristic of deep sea fish—both American and European eels are believed to return to the Sargasso where they spawn and die.

7.5.5 SALMONS AND EELS: COMPARISONS AND CONTRASTS

The anadromous and catadromous migration patterns shown by Pacific salmons and North Atlantic eels respectively represent extreme opposites in tactics, although in both groups the adults spawn once and die. Salmons produce relatively few large demersal eggs in freshwater, the eels many small pelagic eggs in the sea; salmons grow quickly in a productive marine environment, eels grow slowly in a less productive freshwater environment; in salmons mating is restricted to individuals born in the same river or stream—and in pink salmon quite strictly to members of the same generation or year class—whereas in eels the whole mature population may be involved and this will include many year classes. As judged by the abundance of salmon and eels, both tactics are obviously successful. But the capacity to adjust to change would appear to rest on two different approaches: each species of salmon has numerous separate stocks between which there is little or even no mixing; each species of eel has but one stock. So in salmons the capacity to meet change is perhaps assured by a variability of stocks, whereas in eels the variability lies within the population as a whole: the contrast is between species which have developed assortative (selective or restrictive) and panmictic (random) mating strategies respectively.

7.5.6 HERRING

The Atlantic herring, *Clupea harengus*, lays demersal eggs in restricted and

132

well-defined areas and in the Northeast Atlantic there are stocks which spawn in spring, summer, autumn and winter.

The Atlanto-Scandian stock is contained within the subpolar gyral in the Norwegian Sea. This stock spawns in spring off the southwest coast of Norway. After hatching the larvae drift in the north-going current and the young fish spend their first two years in coastal waters. The older and mature fish are oceanic and in summer their main feeding area is along the line of the productive polar front between Atlantic and Arctic waters from Spitsbergen to Northeast Iceland. In the autumn the fish move to the south west along the western edge of the Norwegian Sea gyral and in winter they concentrate in cold water, 3°–4°C, at depths of 300 to 400 m to the north of the Faroes. In March the ripe herring reach the Norwegian coastal spawning grounds and so complete an annual migration circuit of over 3000 km.

There are summer, autumn, and winter spawning stocks in the northern, central and southern North Sea respectively. The pelagic larvae are carried by the residual currents to their nursery grounds. In the North Sea some of the summer-spawned fish grow up in the Scottish Firths while others are carried across the North Sea into the Skagerrak and into Norwegian and Danish waters; some of the survivors of the autumn spawning in the central North Sea grow up in English coastal waters, while others are carried eastwards into the German Bight; similarly, the young stages of winter-spawned herring are carried into English and continental waters bordering the Southern Bight and into the German Bight in years of strong north-easterly drift. On the nursery grounds there is some mixing between fish spawned in different areas and this is probably most marked between fish spawned in the central and southern North Sea. North-Sea herring first spawn when 3 or 4 years old and thereafter the adults of all three stocks share a common feeding ground in the northern North Sea (Fig. 7.6). Nevertheless, there are consistent meristic or numerical differences (for example, in mean vertebral count) between the spawning stocks which indicates some order in their return to the spawning grounds.

7.5.7   TUNA

The albacore, *Thunnus alalunga*, is the most valuable of the tuna (the meat is white) and is fished in Japanese and American waters in the North Pacific. There is probably one stock of albacore in the North Pacific and this is contained within the subtropical gyral (Fig. 7.7). Albacore mature at a length of 90 cm and spawn in the North Equatorial current between the Hawaiian Islands and the Philippines. The eggs are pelagic but nothing is known of the early life history of the young tuna until they appear in the American summer fishery as 2- and 3-year-olds. The American fishery peaks in August and subsequently some of the younger and nearly all the older fish move to the west of the 180° meridian and enter the Japanese fishery which lasts from October to March. The older fish do not return to American waters. The distribution of albacore on both sides of the Pacific appears to be related to the fronts between water masses: in the north they feed in the areas of divergence towards the edge of the subtropical gyral and in the south in the areas of convergence in the equatorial region.

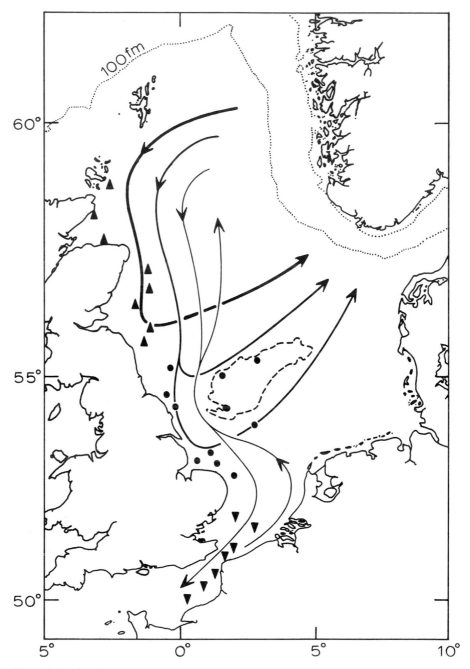

Figure 7.6. Migration routes and spawning grounds of three herring stocks in the North Sea. ▲ Northern North Sea summer spawners. ● Middle North Sea autumn spawners. ▼ Southern Bight and Channel winter spawners. The three groups share a common feeding ground in the northern North Sea. [After Harden Jones 1968]

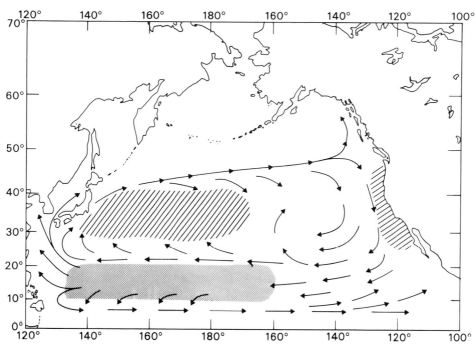

Figure 7.7. Distribution of albacore in the North Pacific. The areas of the American fishery on young albacore and the Japanese fishery on older albacore are cross-hatched, and the area in which the tuna are believed to spawn is stippled. The main features of the North Pacific subtropical gyre (Kuroshio, Kuroshio extension, California, California extension and North Equatorial currents) and the North Equatorial counter current are shown.

7.5.8   HAKE

Merluccius productus, the North Pacific hake, ranges from Alaska to Mexico, and it is one of the most abundant fish associated with the California current and its upwelling system. The hake leads a pelagic life over the continental shelf (Alverson & Larkins 1969). The young and immature fish (less than 50 cm in length and up to 4 years old) are distributed over the southern part of the range. In summer the older fish (more than 5 years old and over 50 cm in length) feed in more northerly waters. In the winter the maturing adults move off the shelf and migrate southwards to their spawning area which extends over deep water about 300–400 km off the coast of California (Fig. 7.8). After spawning the spent fish return to the north and shallower waters.

The eggs are pelagic, and while the distribution of the larval stages has been described, little is known of the distribution or movements of the 1- to 3-year-old hake.

7.5.9   COD

The Atlantic cod, *Gadus morhua*, is widely distributed in the North Atlantic: the existence of several independent stocks has already been mentioned. The Arcto-Norwegian stock feeds on shallow banks in the Barents Sea. During the summer and in the autumn the cod move to winter areas in deeper water near

and related behaviour are linked in time to seasonal and other changes in the environment.

Elvers of the European eel respond positively to a chemical in inland water to which they have never previously been exposed: the reaction would appear to be innate. Salmonids recognise their parent stream or tributary by reference to a chemical signature—detected through the nose as a smell—to which they were conditioned, or imprinted, when they were in freshwater. In general the chemical clue probably functions as a sign stimulus and releases a positive response to a water current (swimming upstream) and so combines long distance detection by the most sensitive of the chemical sense organs with an unambiguous directional response.

However, very little is known of the clues used when on migration between, for example, a feeding area and a spawning area. Suggestions include features of bottom topography for animals that are close to the bottom and celestial clues or coastal landmarks for animals that swim close to the surface or stick their heads out and look round (gray whale); water currents and magnetic fields are other contenders. There are certainly problems of directionality; even fish whose migration circuits are essentially downstream may move at ground speeds which are too fast to be accounted for by passive drift: the migrants may be swimming, and therefore oriented, downstream. This may also be true for fish which use tidal transport in shelf seas.

On a day-to-day, or even hour-to-hour basis, the fine control, or triggering, of migratory and spawning behaviour, may be related to lunar periodicity and tidal cycles. This seems very likely in shelf seas. But in temperate waters little is known of such matters and the point has already been made that lunar periodicity is, at present, best documented in reef fish. At a seasonal level, feeding, growth, gonad maturation and migration must be linked to the production cycles. In subtropical, temperate and polar waters the seasonal changes in daylength, possibly detected through an extraocular photoreceptor—photosensitive vesicles in cephalopods, pineal gland in fish—may provide the link with the endocrine system. In this way the seasonal changes in solar radiation, which drive the production cycles, would also initiate and in part control the movements and migrations of the nekton which so depend upon them.

142

8

Utilization of Aquatic Production by Man

W.E. ODUM

8.1 Introduction

A commonly repeated refrain is that the oceans have the potential to provide much of the future food needs of the world's burgeoning human population. In this chapter we will critically examine this idea by looking first at the present harvest of aquatic organisms by man and then speculating about the probable upper limits of future yields. A further point to be investigated is the relative importance of the present aquatic yield compared to that from terrestrial sources.

8.1.1 THE CURRENT AQUATIC HARVEST

In 1977 the yield of aquatic organisms to man from fisheries and aquaculture was estimated by F.A.O. to be approximately 73.5 million tonnes (metric tons, wet wt; see Table 7.1). About 86% of this originated from oceans and estuaries and 14% from freshwater. Almost all of the harvest consisted of fish and shellfish with a little more than 1 million metric tons of red and green macroalgae.

Surprisingly, only about 70% of the aquatic harvest is actually consumed by human beings; the remainder is reduced to fish meal and oil. Most of this, in turn, is fed to domestic animals such as chickens, pigs, cattle, dogs, cats, and even pond fishes.

When compared to other sources of human food on a dry weight basis (Table 8.1), the aquatic harvest forms an insignificant fraction. Even when compared on the basis of energy content (kcals), aquatic foods contribute only 0.5–1.0% of the energy in man's diet. These comparisons, however, can be highly misleading. When compared in terms of total protein, fish and shellfish provide between 5 and 6% of the total world human consumption of protein and almost 15% of the animal protein in our diet. Several additional percentage of indirect consumption are provided through fish meal and oil fed to domestic animals which are then eaten by man.

As you might expect, direct consumption of fish and shellfish varies greatly from country to country. Although the world average is about 12 kg (wet wt.) per person per year, it ranges from less than 0.05 kg in Afghanistan to over 40 kg in Iceland. In locations such as Japan, Iceland, and many of the Scandinavian countries which have had a long, historical dependence on fish and fish products, more than 25% of all dietary protein comes from fish. In the developing countries of Asia, Africa, and Central and South America, animal

Table 8.1. Approximate harvest of human foods (dry wts. in millions of metric tons).

From terrestrial sources		
Cereal grains		1000
Other food crops		220
Milk		48
Meat		20
Eggs		5
	Total	1293
From aquatic sources		
Fish and shellfish		11
Marine plants (macroalgae)		0.2*
	Total	11.2

*Much of this goes into chemical products such as medicines and vegetable gums.

protein is so scarce that fishery products provide an average of 20% of all animal protein consumed by humans. Bangladesh, an extreme example, derives 80% of its animal protein from fish, largely produced in aquaculture ponds.

For certain countries, fishery products are an important economic consideration. As examples, Iceland, Japan, Canada, Denmark, and Peru (prior to the collapse of the anchovetta fishery), all have a critical dependence for foreign exchange based on fishery exports. Fully 90% of the value of all Icelandic exports consists of processed fish.

8.1.2 A COMPARISON OF AQUATIC AND TERRESTRIAL HARVESTS

Thus far we have established that the land provides most of man's food needs while the only major contribution from the aquatic harvest is animal protein. With this in mind, one would expect to find fundamental differences between the patterns of harvest from land and water. They are, in fact, virtual mirror images of each other (Fig. 8.1).

On one hand, terrestrial agricultural yields are predominantly plant products such as cereal grains; herbivores compose a small percentage while the harvest

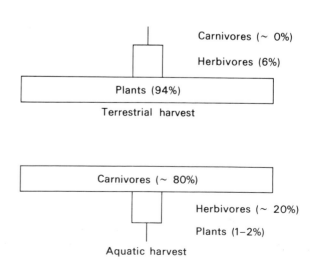

Fig. 8.1. A comparison of world harvests from aquatic and terrestrial sources. [After Whittaker & Likens 1975, and F.A.O., 1968]

144

of carnivores is essentially nil. When was the last time you ate a tiger steak or a wolf burger? The only exception to this pattern is the fraction of chicken, hog, and cattle production which results from the consumption of relatively small amounts of fish meal contained in processed foods.

In stark contrast, the aquatic harvest is composed of a tiny fraction of plant material, a moderate amount of herbivores (oysters, clams, mussels), and is dominated by carnivores. Tuna, salmon, and cod for example, function energetically between the third and fifth trophic levels.

This contrasting harvest pattern cannot be attributed to significant differences in either primary or secondary production between land and water (Table 8.2). It can, however, be traced to major differences in harvest efficiency.

Terrestrial harvest efficiency, although very low (Table 8.2), is 30–40 times higher than aquatic harvest efficiency. On land the relatively few plants which

Table 8.2. A comparison of characteristics between land and water. All weights are dry weights. [Revised from F.A.O., 1977 and Whittaker and Likens 1975]

	Land	Water
Total area	145×10^6 km^2	365×10^6 km^2
Energy conversion efficiency (sunlight/fixed carbon)	~0.13%	~0.03%
Total net 1° production	$90-120 \times 10^9$ tons/year	$50-60 \times 10^9$ tons/year
Total 2° production ~(herbivores only)	~0.8×10^9 tons/year	~3.0×10^9 tons/year
Total harvest by man (plants and animals)	~1.3×10^9 tons/year	~17.0×10^6 tons/year
Harvest efficiency (total harvest/net 1° prod.)	~1.0%	~0.03%

are suitable human food are concentrated in small areas, nurtured carefully, and harvested efficiently. Herbivorous animals are managed the same way. Carnivores are not feasible to raise due to the tremendous energy loss which accompanies trophic transfer.

In aquatic ecosystems the majority of plants and animals are potential human food. Plants, however, are generally microscopic and thoroughly dispersed while aquatic animals are so scattered, occur at such low concentrations, and are commonly so mobile that it is difficult to harvest them efficiently. The best strategy has been to ignore plants and concentrate upon the upper portion of the food web which consists of relatively large finfish and shellfish.

Further contributing to the problem of harvesting the ocean are factors such as unpredictable and dangerous sea and weather conditions, unforeseen movements of entire fish stocks, and the sociological and economic difficulties associated with sending fishermen to sea.

The extreme inefficiency of most types of aquatic harvesting becomes even more apparent when net energy calculations of land and water harvesting are compared (Fig. 8.2). Even though terrestrial farming requires extensive energy subsidies in the form of fertilizers, pesticides, irrigation, and cultivation, the energy produced as crops usually exceeds the amount of energy required to grow them. This is not the case with husbandry of cattle and chickens nor with

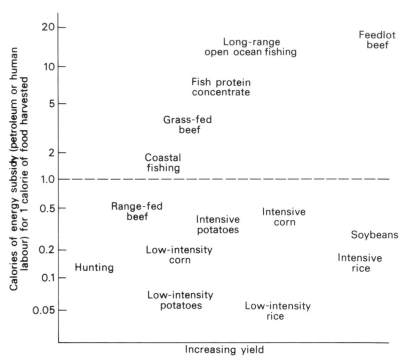

Fig. 8.2. The relationship between energy inputs in the form of petroleum and human energy subsidies, and energy obtained from the food product. [After Steinhart & Steinhart 1974]

aquatic harvests. Fishing, especially long distance trips on the high seas, consumes more energy in the form of fuels, boat construction, and boat maintenance than is provided by the catch. In the past this has not been of much importance due to the ready availability of cheap fossil fuels. In the future, however, this may become of critical importance.

If future aquatic harvests are to be increased dramatically, ways must be found to circumvent the problems which we have discussed. These may include:

1 Harvest of organisms at lower trophic levels, and
2 Creation of techniques to concentrate organisms more efficiently and which approach a net energy balance.

To determine whether these are realistic possibilities, our next step will be to examine both conventional fisheries and aquaculture as they are currently practised.

8.2   World fisheries

In examining the world fishing industry, several important questions must be addressed:

1 What is the technological state of the fishing industry?
2 Who are the leading fishing nations and to what factors do they owe their pre-eminence?

146

3 What areas of the world and which species are most important?
4 Can and should the industry be made more efficient?
5 Is there an upper limit to the world fish catch?

8.2.1 TECHNOLOGICAL STATE—HUNTING

The single common feature of all fisheries, whether simple, freshwater pond harvesting or technologically advanced open ocean trawling, is that they are basically hunting operations. Even with such refined equipment as modern, high speed vessels, satellite navigation, advanced side-scanning sonar, synthetic nets, and spotter aircraft, the fishermen are dealing with a mobile resource which due to ecological or oceanographic vagaries may or may not be available to the fishermen. To a certain degree, aquatic harvesting, with the exception of controlled aquaculture, remains at a level which terrestrial harvesting surpassed several thousand years ago.

Because of environmental and biological uncertainties, the yields of many fisheries tend to fluctuate over short periods of time. For example, the Peruvian anchoveta (*Engraulis ringens*) fishery which had reached an annual catch of over 13 million metric tons by the early 1970s suddenly plummeted to less than 3 million metric tons by the mid 1970s. This collapse was evidently precipitated by meteorological and oceanographic changes and further enhanced by overfishing. Many other fisheries show marked year-to-year fluctuations in catches due to a variety of factors including:

1 Water temperature changes.
2 Changes in larval survival.
3 Alterations of migration pathways.
4 Pollution events.
5 Overharvesting by man.
6 Adverse weather conditions for the fishermen.

8.2.2 THE LONG-TERM TREND IN THE WORLD AQUATIC HARVEST

As shown in Fig. 8.3, the world harvest of aquatic organisms rose steadily from the end of the Second World War until the early 1970s. This pattern of growth was due to a combination of factors including development of technologically advanced fishing vessels, advances in fishing gear and fishing strategy, expansion to new fishing grounds, the extension of fishing effort to species which formerly were not harvested and, above all, an increased demand and higher market value for fish.

The levelling off of the world fish catch during the 1970s, which may be temporary, can be attributed to the impact of rising fuel costs, unusual declines in major fisheries such as the Peruvian anchovetta and the North Atlantic haddock, and, perhaps, the first effects of the move toward 'enclosure' or the adoption of 200 mile fisheries jurisdictions.

8.2.3 THE MAJOR FISHING COUNTRIES

In 1976, 114 countries landed 10 000 metric tons or more of aquatic organisms.

Fig. 8.3. The total world harvest of aquatic organisms excluding whales. [From F.A.O. Fisheries Statistics yearbooks 1948–1976]

However, of the world harvest of 73.4 million metric tons, the combined landings of six nations accounted for more than one half of the total. Even more interesting is the fact that the leading three fishing nations accounted for 38% of the total harvest. In short, world fisheries are dominated by a few nations, leaving the majority with a small slice of the world fisheries pie.

The six leading fishing nations are shown in Table 8.3. The factors responsible for their elevated positions are as varied as the countries themselves. Much of the impetus for the rapid increases in the world catch during the 1950s and 1960s came from the remarkable expansion of the fishing fleets of the U.S.S.R. and Japan. Both countries adopted national policies of government subsidization and strong encouragement of their respective fishing industries in response to projections of increasing demand for fish protein. The result was an industry based on modern, technologically advanced fleets of stern trawlers and large factory ships capable of remaining at sea for periods of many months and fishing for unutilized and underutilized species in remote locations. A further result has been a steady and rapid increase of the annual catch of both countries through the mid 1970s.

Table 8.3. The six leading fishing nations in 1976. Catches are in millions of metric tons. [From F.A.O., 1977]

Japan	10.6
U.S.S.R.	10.1
Republic of China	6.9
Peru	4.3
Norway	3.4
U.S.A.	3.0

The reasons behind mainland China's high position as a fishing nation are totally different from the other leading countries. China relies on the combined catch of a huge fleet of fishing junks and similar small vessels which intensively fish the country's extensive and productive territorial waters. Further augmenting the catch is the yield of literally millions of freshwater and brackish water aquaculture ponds and flooded rice paddies. These ponds usually are farmed for various species of carp and are fertilized with human and animal manures along with unutilized portions of plant crops.

The fishing industry of Peru is based almost exclusively on a single species, the anchovetta. As a result the country's success as a fishing nation fluctuates along with the abundance of the anchovetta. During the late 1960s Peru was the world's leading fishing nation. In recent years the catch has dropped to a few million metric tons and threatens to go even lower.

Norway has a traditional dependence upon fish and fish products. This has created a steady demand which, in turn, has enabled Norway to modernize and expand its fishing fleet. This has culminated in a fish catch which has increased each year and enabled Norway to replace the United States in fifth place.

The United States' relatively high position is based on extensive, highly productive coastal waters and a long-established, but static, fishing industry. Since the Second World War the U.S. fishing industry has not grown markedly or experienced much modernization of its fleet except in certain, high value fisheries such as tuna and shrimp. The result has been that the total catch of the United States between 1950 and 1970 remained virtually the same. Since both the *per capita* consumption of fish and the human population grew significantly during this period, the United States has depended more and more on imports from countries such as Japan, Canada, and Iceland.

The relative positions of the remaining fishing nations of the world can be attributed to a variety of factors. Some, like Japan and the U.S.S.R. have increased their catches dramatically through heavy financial investment and modernization of their fleets. Others, similar to the United States, have failed to modernize or expand their fleets resulting in stable or even declining catches. Still others are limited by small territorial seas, little or no capital for investment, lack of efficient distribution systems for easily spoiled fishery products, or simply a low demand for fish. The recent enactment of 200-mile fishery zones by many countries has been an attempt not only to establish better management practices but also to offset the technological advantages of the more advanced fishing countries. Within a few years the rankings of the countries relative to each other will probably experience some readjustment.

While analysing the total catches of the different countries, it is important to remember that weight alone is often a poor indicator of the importance of the catch. For example, a ton of Peruvian anchovettas might typically generate a dockside price of U.S. $40 while a ton of Australian shrimp and prawns might be worth U.S. $1000. At the time that Peru was the world's leading fishing nation in terms of total weight landed, it was not in the top ten in terms of value of the catch.

Due to increasing demand, it is possible for the value of a catch to increase substantially from year to year even though the weight of the catch remains the same. In the United States between 1963 and 1973, while the weight of the catch changed little, its total value increased by 140%. Inflation accounted for less than one third of this increase.

8.2.4 ECOLOGICAL, TAXONOMIC, AND GEOGRAPHICAL COMPOSITION OF THE CATCH

The approximate contributions of different groups of aquatic organisms to the total world aquatic harvest are shown in Table 8.4. Although the species groups as devised by F.A.O. for their book-keeping purposes are taxonomically and ecologically vague, there are several interesting points which emerge from this table. First, as pointed out previously, aquatic plant harvests are extremely minor in comparison with animal harvests. Second, invertebrates such as molluscs and crustaceans are also relatively minor in terms of weight landed although their relative economic value is much higher. Third, most of the fish catch consists of demersal and benthic species such as cod (*Gadus morhua*), haddock (*Melanogrammus aeglefinus*) and flounder (family Pleuronectidae); exceptions include the pelagic fishes such as herring and sardines (family Clupeidae), anchovies (family Engraulidae), tuna (family Scombridae), and billfishes (family Istiophoridae). Finally, roughly 80% of the catch of fishery organisms are first, second, and third order carnivores. Less than 20% of the catch is composed of herbivores such as oysters (family Ostreidae), mullet (*Mugil* spp.), and milkfish (*Chanos chanos*).

Table 8.4. Principal groups of aquatic organisms contained in the world catch for 1974. Mammals are excluded. [F.A.O., 1975]

Species group	Catch (million metric tons)
Herring, sardines, anchovies *et al.*	13.7
Cod, hake, haddock *et al.*	12.7
Miscellaneous marine and diadromous fishes	9.1
Freshwater fishes	9.0
Redfish, sea bass, conger *et al.*	4.6
Mackerel, snook, cutlassfish *et al.*	3.6
Molluscs	3.4
Jacks, mullet, sauries *et al.*	3.3
Salmon, trout, smelt *et al.*	2.4
Crustaceans	1.9
Tuna, bonito, billfish *et al.*	1.9
Miscellaneous aquatic plants and animals	1.5
Flounder, halibut, sole *et al.*	1.2
Shad, milkfish *et al.*	0.7
Sharks, rays, chimaeras *et al.*	0.6
Rounded total	70.0

The geographical origins of the world fish catch are shown in Table 8.5; the unevenness of the catch is obvious. Two areas, the N.E. Atlantic and the N.W. Pacific account for almost half of the total catch. Other areas such as the Mediterranean, Black, and Arctic seas have insignificant yields. This contrasting pattern is due to at least two factors:

1 The relative magnitude of biological productivity of the area.
2 The ease of access to the fish stocks by fishermen.

Table 8.5. Total world landings (conventional fisheries and aquaculture) by geographical area. Values are in million metric tons (mmt). Because of rounding errors, totals may not always equal sum of sub divisions. [F.A.O., 1977]

Area		1965	1976
Atlantic			
N.W. Atlantic		4.0	3.5
N.E. Atlantic		9.6	13.3
Central Atlantic		2.6	5.1
Mediterranean and Black Seas		1.0	1.3
S.W. Atlantic		0.5	1.2
S.E. Atlantic		2.2	2.9
	Atlantic total	19.9	27.3
Pacific			
N.W. Pacific		10.7	17.2
N.E. Pacific		1.5	2.4
Central Pacific		3.3	6.9
S.W. Pacific		0.1	3.8
S.E. Pacific		8.2	5.6
	Pacific total	23.8	32.6
Indian Ocean		1.9	3.3
Antarctic Sea		—	0.001
Arctic Sea		—	—
Freshwater		7.6	10.3
	World total	53.2	73.4

The three areas with the highest yields either have vigorous oceanic upwelling in close proximity to land or shallow banks with local upwelling. All three are intensively fished. In contrast, the Mediterranean and Black seas, although easily accessed by fishermen, have relatively low biological productivity due to a combination of factors including lack of extensive upwelling.

If the major world oceans are analysed in terms of fish yield per unit area, we find that in 1976 the Atlantic produced roughly $257 \, kg/km^2$, the Pacific $181 \, kg/km^2$, and the Indian Ocean only $44 \, kg/km^2$. Although many areas of the Indian Ocean have high levels of biological productivity, the countries in closest proximity, such as India, have been unable to exploit these resources except in shallow coastal waters. The Indian Ocean appears to be the greatest underutilized fishery resource remaining in the world.

John Ryther, an American oceanographer, has pointed out that most of the potential yield from the oceans is found in a relatively small percentage of the total area (see Chapter 7 and Table 7.2). By his estimate, the open ocean, which comprises 90% of the surface of the ocean and an even greater volume, holds less than 1% of the harvestable fish. Coastal waters, which cover about 10% of the total area of the oceans, support almost 50% of the potential fish catch, and areas of oceanic upwelling, which cover only about 0.1% of the area, have the remaining 50% of potential fish stocks. While these estimates are only rough approximations, it is safe to conclude that 10% of the area of the world's oceans holds almost all of the potential yield and that the open oceans beyond the continental shelf are virtual deserts from the standpoint of fisheries yield to man.

8.2.5 THE PHILOSOPHY OF FISHERIES MANAGEMENT AND THE CONCEPT OF MAXIMUM SUSTAINED YIELD

The prevalent fisheries philosophy prior to the twentieth century was based on the assumption that fish populations in oceans and large bodies of freshwater were virtually inexhaustible. By the turn of the century, however, it was becoming increasingly noticeable that the catches of certain heavily fished populations were not increasing in response to increasing fishing pressure. This was particularly true for restricted areas such as the Mediterranean and North Sea; for example, introduction of new, steam powered fishing vessels in the plaice and herring fisheries of the North Sea resulted in only minor increases in the total catch.

Gradually, as the same pattern of stable or even declining yields in response to increased fishing pressure was repeated elsewhere with other fish species (Fig. 8.4), it became apparent that all fisheries have an upper limit of exploitation

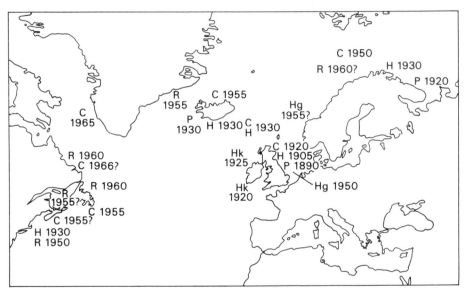

Fig. 8.4. Overfishing in the North Atlantic. The years are the approximate dates by which fishing on the stocks indicated reached a level beyond which further increases in fishing effort gave no sustained increase in the catch. C, cod; H, haddock; P, plaice; R, redfish; Hk, hake; Hg, herring. [From F.A.O. 1968]

and can be overfished. To understand and predict the consequences of over-fishing, fisheries scientists have developed a series of models based on the population dynamics of specific fish stocks. With these models it is possible to incorporate catch data with species-specific biological data and produce a model which has reasonable predictive ability. Central to many of these models is the concept of maximum sustained yield (MSY) which is based on long term estimates of fishing effort, yield, and the biological characteristics of a particular fish stock such as George's Bank haddock.

An oversimplified but convenient way to depict the concept of maximum sustained yield is with a catch-effort curve (Fig. 8.5). This curve is usually applied

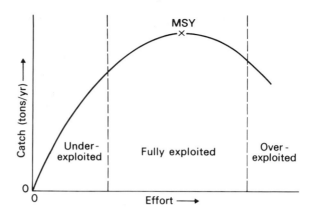

Fig. 8.5. Catch/effort curve showing the relationship between increasing fishing effort and the resulting catch. Fishing effort may be estimated in a variety of ways including number of ships or pieces of standard gear; all are considered on the basis of a standard unit of time such as a year. The exact shape of the curve depends on the growth characteristics of the fish population. This particular curve assumes logistic growth.

to a single stock of fish and includes certain assumptions such as constant natural mortality, constant recruitment, and density independent growth. Unfortunately, these constraints can rarely be satisfied in the real world. In addition, this approach ignores most aspects of ecosystem level interactions and processes. Finally, there are many cases which, for economic, biological, or sociological reasons, may have an optimum sustained yield (OSY) at a lower level of effort than the MSY. For example, consumers or sports fishermen may require that individual fish be above a certain size. To obtain large numbers of fish of this size from the population, it may be necessary to lower the effort and thus allow the average individual fish a longer growth period in order to reach the desired size.

There are a multitude of arguments regarding the usefulness of the MSY concept, its appropriateness for specific populations, and whether it should be applied in terms of biomass or economic value. Certainly, it is difficult to apply to fisheries in which natural mortality or larval recruitment vary significantly from year to year. In spite of these problems, MSY has been used extensively for the past 25 years as the management basis for many fisheries. In the absence of better tools, this approach offers a useful and general measure of the state of a fishery and the likelihood of future yield increases.

8.2.6   POTENTIAL FUTURE HARVESTS—WHAT IS THE UPPER LIMIT?

When the world's conventional fisheries are examined in respect to MSY, we find, surprisingly, that most are nearing MSY, have reached MSY, or are overexploited. Overexploitation often has occurred because wild fish stocks are common property resources and can be exploited by anyone. In dealing with common property resources, the individual rarely behaves in terms of the group optimum, but concentrates instead upon short term gains. This has been complicated further by the fact that many of the world's important fishing grounds are located in international waters. As a result as many as 15 to 20 countries with vastly different economic and, even, scientific philosophies may haphazardly exploit a single fish stock with little regard for the long term effects. The decline of catches of the blue whale (*Balaenoptera musculus*) and cod (*Gadus morhua*) provide two good examples.

153

Considering the fact that many individual fisheries are fully exploited, one wonders whether the total yield from the oceans is also approaching an upper limit. During the past 20 years there have been a number of attempts by scientists to answer this question from a theoretical point of view using two basic approaches. The first, which might be called the 'food web approach', consists of:

1 Estimating net primary production for all sections of the world's oceans.
2 Identifying the average trophic position of fishery organisms from those sections.
3 Estimating the secondary production of fishery organisms at that trophic level by converting from the net primary production estimate with food chain transfer efficiencies.
4 Estimating what percentage of the secondary production of fishery organisms can feasibly be harvested by man.

By summing estimates from all sections of the oceans, it is possible to obtain an estimate for potential world annual yield.

The second method, an extrapolation technique, consists of three steps. First, all conventional and potential fisheries are identified. Second, the potential MSY from each fishery is estimated. Finally, all of these estimates are added together to obtain a potential estimate for the world.

The early estimates using both techniques made in the 1950s and early 1960s tended to cover a great range of values. In recent years, however, the basic data base has improved considerably and most estimates have tended to cluster around an upper level of 100–120 million metric tons (mmt).

There are, of course, many reasons why this potential upper limit may be too conservative. Presently employed food-chain transfer efficiencies may be too low. These efficiencies are usually based on transfer between two species; transfer through ecosystems may be more efficient. Furthermore, in the future it may be possible to harvest organisms at lower trophic levels and to harvest species which are not harvested now. The latter case is a strong possibility if fish protein concentrate (FPC) becomes widely utilized. FPC is a clean, white powder which has a high protein content and can be manufactured from almost any part of any fish. Its wider acceptance in the future could lead to significant increases in the catch of species which are currently regarded as inedible.

The antarctic krill (*Euphausia superba*), a coldwater euphausid not unlike a small shrimp, is a low trophic level organism with much unrealized potential. Krill occur in the Southern Ocean near the Antarctic ice shelf in enormous numbers. They are harvested by Chile and the U.S.S.R., but only to a limited extent. Although relatively easy to harvest, they spoil quickly and must be processed rapidly into a frozen product or canned paste. The MSY of krill is totally speculative but could be as high as 50–100 mmt. If either FPC or krill become important industries and are accepted as human food, they have the potential to double the current estimates of the maximum world fisheries yield.

All of these possibilities must be considered remote at the moment, particularly in light of the rapidly increasing costs of fossil fuels. We are forced to

154

conclude that 100–120 mmt is the most likely maximum yield from the world's conventional fisheries.

8.3 Aquaculture

Since the world fish catch appears to be approaching an upper limit, at least for the foreseeable future, interest recently has become focused more and more on controlled 'fish farming' or aquaculture. Although there are many inherent biological, technological, and socioeconomic problems associated with aquaculture, the potential for substantially increased production is very encouraging. The ultimate objective, of course, is to create a significant food producing industry, analogous to agriculture, with a reasonably predictable and controllable yield.

8.3.1 THE PRESENT STATUS OF AQUACULTURE

Although statistical records for aquaculture are sketchy, particularly for finfish, it is possible to generate a rough estimate of world aquaculture production (Table 8.6). This value of 6 million metric tons (wet wt.) is less than 10% of the

Table 8.6. Estimated world aquaculture yield for 1975. [From Pillay 1976]

Organisms	Million metric tons (wet wt.)
Finfish	4–5
Seaweeds	1.1
Oysters	0.6
Mussels	0.2
Scallops	0.06
Clams	0.04
Cockles and other molluscs	0.03
Shrimps and prawns	0.02
Total	≈6.0

estimated world fish catch for the same year, but represents over a 50% increase since 1970. Several countries including Japan, Poland, and Romania have experienced two to five-fold increases in production in this 5-year period. In other words, although the present yield from aquaculture is relatively small, considerable growth potential exists.

Although more than 100 species of plants and animals are cultured, greater than 60% of the total aquaculture yield comes from pond culture of a few species of finfish. These include rainbow trout (*Salmo gairdneri*), salmon (family Salmonidae), channel catfish (*Ictalurus furcatus*), mullet (*Mugil* spp.), milkfish (*Chanos chanos*), and several species of carp (family Cyprinidae) and tilapia (*Sarotherodon* spp.).

Other than pond culture of fishes, other important types of aquaculture include:

155

1 Culture of oysters (family Ostreidae) and mussels (family Mytilidae) either on the sea bed or suspended from rafts.
2 Farming of scallops (family Pectinidae), clams (families Veneridae, Solenidae, Myacidae, etc.) and cockles (family Cardiidae) on controlled areas of seabed.
3 Culture of fishes in floating cages.
4 Pond culture of shrimp (family Penaeidae) and freshwater prawns (*Macrobrachium* spp.).
5 Abalone (*Haliotis* spp.) culture.
6 Culture of marine algae or 'seaweeds'.
7 Culture operations based on minor species such as eels and turtles.

Almost all aquaculture products are intended for direct human consumption, but there are exceptions such as the Japanese cultured pearl industry. Approximately 30 genera of large marine algae, popularly known as 'seaweeds', are grown commercially, largely for industrial purposes. Certain of the red algae (primarily species of *Porphyra*) are raised for human consumption by the Japanese. This 'nori' culture produces annually about 120 000 wet metric tons of high quality food (30–50% protein on a dry weight basis). However, the majority of red, green, and brown macroalgae which are harvested or cultured by man contain pentosans (five-carbon carbohydrates), compounds which are not easily digested by mammals. Instead of food, these algae are used as a source of vegetable gums (alginates and agar-agar) which are included in a range of products including pharmaceuticals, paper, adhesives, textile sizing, and synthetic rubber.

8.3.2 GEOGRAPHICAL SOURCES OF AQUACULTURE PRODUCTION

Almost 75% of the total world aquaculture yield comes from the Indo-Pacific region, an area bounded roughly by Pakistan on the west, Japan and the Philippines on the east and Australia to the south. The traditional ponds and flooded rice fields of the People's Republic of China produce at least a third of the total world production, often in ponds that have been operated for hundreds of years. Other important aquaculture countries in this region are Japan, Korea, Taiwan, the Philippines, and Indonesia.

In contrast, the development of aquaculture in most western countries has remained sluggish due to low market demands and extreme competition and high costs for potential sites for aquaculture installations. In the United States, for example, aquaculture contributes only a tiny fraction of total meat consumption and then only in the form of high-priced speciality items such as trout, crayfish, and catfish. There are countries (e.g. Israel) and specific industries (e.g. mussel culture in Belgium, France and Spain) of more than minor importance, but the general picture at this time in western Europe, Africa, and North and South America is one of a minor industry with unrealized potential.

8.3.3 CONTEMPORARY AQUACULTURE TECHNIQUES

Basically, two strategies are employed by the nations which have significant aquaculture production. One approach, found principally in nations which have a high human population and a scarcity of animal protein, is to grow species such as carp and mullet which occupy relatively low trophic positions and grow rapidly in small, confined areas, but which may not have a high market price. These nations optimize for large quantities of animal protein in preference to substantial economic profits.

The second strategy is to grow lesser amounts of high quality, expensive products such as trout and shrimp. These organisms often are carnivorous and usually require more expentive care (e.g. clean water, low densities, and water with a high oxygen content). This strategy is used most often in developed countries such as the United States and Japan, although a few developing countries have experimented with this approach to aquaculture in hopes of generating valuable export products.

From a technical standpoint, contemporary aquaculture techniques range from very simple manipulations of the natural environment with little protection of the culture organisms to extremely elaborate artificial environments with total control and protection of the organisms. The simplest procedures, referred to as 'extensive' operations may consist merely of planting seed oysters or clams on a suitable area of sea bed and then returning several years later to harvest what remains. Incursions of predators, lack of food, or destruction of oysters by water quality changes cannot be avoided.

More control can be obtained with a moderately 'intensive' procedure such as the raft culture of oysters. In this case, almost total protection is afforded from benthic predators and the raft can be moved to find more suitable feeding conditions, better water quality, or simply to avoid storms. Complete control can be achieved through a highly intensive operation such as tank or bottle culture of oysters. This allows total control of water temperature, water quality, and food availability; moreover, predators can be completely excluded and diseases and parasites controlled rapidly and effectively.

Obviously, as aquaculture becomes more intensive the potential yield per unit volume rises dramatically. Unfortunately, the energy subsidies and costs also rise rapidly with the result that few truly intensive aquaculture facilities exist except as laboratory test models. Since energy and technological costs are rising more rapidly than fish prices, it is likely that most truly intensive aquaculture technology will remain on the drawing board for the near future. On the other hand, moderately intensive facilities such as raft, cage, and pond culture, have immediate potential and will probably form the basis for increases in aquaculture production for the next 10–25 years.

8.3.4   IDEAL CULTURE ORGANISMS

Although a wide range of marine and freshwater animals and plants have been tried as culture organisms, relatively few have proved to be both a biological and economic success. This is because most aquatic organisms have one or more characteristics which make them unsuitable for aquaculture. Ideally, a culture organism should have a simple life history and have the capability of being reared through all life history stages in captivity. If the life history is too complex and contains too many developmental stages with strict water quality and food requirements, then the time and money involved in rearing may make the total operation not feasible. There are species such as the milkfish which have not been spawned in commercial quantities in captivity but still form the basis of a substantial industry. These, however, are the exception to the rule and their culture is always plagued by uncertainties associated with an irregular supply of juveniles from natural ecosystems. Most aquaculture species such as carp, trout, and oysters can be spawned in captivity and raised with relative ease through several simple larval and juvenile stages.

A second requirement for a culture organism is that it be hardy, resistant to disease, and adaptable to crowding. Some potentially useful species are unable to withstand the disease, parasites, and degraded water quality which usually accompany aquaculture. Others, such as lobsters and crabs, do not react well to crowding and tend to be cannibalistic.

Finally, the trophic position of potential culture organisms is of great importance. In most cases the species should function at a low trophic level (ideally a herbivore or detritivore) so that the most biomass can be produced per unit area without using expensive, animal protein based feeds. For a strict carnivore to be an economic success, either a cheap source of meat must be available (e.g. unmarketable 'trash fish' from a trawl fishery) or the final product must bring a very high market price.

8.3.5 EFFICIENCY OF AQUACULTURE

A comparison of fish farming with terrestrial farming of domestic animals shows that fish can usually be raised more efficiently. Consider a moderately intensive aquaculture operation in which the organisms are fed supplemental prepared fish feeds (e.g. pond culture of trout). In this situation feed conversion rates (ecological growth efficiency) for fish are about one-and-a-half times as great as for swine and chicken farming and almost twice as high as the conversion rates of cattle and sheep.

There are at least two reasons for this greater efficiency. First, because fish are cold blooded they are not forced to expend energy to maintain a constant body temperature. Second, because they are supported by water rather than air they do not need to invest as much energy in a heavy skeletal system or in constantly fighting the force of gravity. Energy savings from both factors can be incorporated into more efficient growth and more rapid weight gain.

For situations in which there is no supplemental feeding, aquaculture usually provides higher yields per unit area than agriculture. In addition to the two previously discussed factors, aquaculture can potentially utilize three dimensions (the entire water column) instead of the two dimensions of a pasture or field. Annual yields of fish as high as several thousand kilograms per hectare are entirely possible while the usual upper limit for cattle is 500–700 kg/ha/year.

In terms of net energy (energy input by man/energy yield to man), aquaculture ranges from values less than one for extensive operations such as sea bed oyster culture to very high values (5–10) for highly intensive culture with heavy energy subsidies as are found in Japanese tank culture of shrimp. As a general rule, net energy values are low for the low trophic position, economically inexpensive fishes such as mullet and carp, and high for the carnivorous, economically valuable species such as catfish and trout.

158

T.V.R. Pillay, an F.A.O. aquaculture expert, predicted in 1976 that world production from aquaculture will probably double to 12 mmt by 1985 and could increase five- to ten-fold in three decades. While the 1985 prediction seems possible, the longer range estimate may be too optimistic.

Certainly, dramatic future increases in aquaculture yields can be accomplished with present-day technology through:

1   Devoting more area to culture.
2   Utilizing more herbivorous species.
3   Employing polyculture or the farming of several species with different trophic positions in the same enclosures.
4   Using greater densities of organisms.
5   Increasing feeding and breeding efficiencies through genetic selection.

On the other hand, a number of factors exist which may limit future production:

1   There are limited areas for expansion, particularly in the 'developed' countries where other uses such as recreation and industry are actively competing for the same sites.
2   Increased water pollution can severely harm the aquaculture industry. This has been a serious problem in recent years in Japan where production in sites around the inland sea has been seriously affected by pollutants.
3   Future shortages and increased costs of fertilizers may be a serious problem.
4   Increased costs of aquaculture feeds may drive the costs of the final product too high.
5   Increased costs of other energy subsidies including petroleum may cause many types of aquaculture to become impractical.
6   Occasional, indiscriminate application of environmental protection regulations designed for other industries may prevent or limit aquaculture. This has occurred in the United States where aquaculture ponds have been considered in the same category as sewage treatment plants and required to control releases of nutrient and high BOD-laden waters. While this may be necessary, it still poses major problems for the aquaculture industry in developed countries.

Probably, the most serious of these limiting factors is competition for suitable space and the increasing costs of petroleum. Still, aquaculture shows definite signs of growing at a greater rate than conventional fisheries. There are large areas, both coastal and inland, particularly in the tropics, which have relatively low natural productivity, no other obvious uses, and which could be used for moderately intensive pond culture of species such as carp, mullet, and milkfish. If these areas are developed in a rational manner, then the possibility of reaching T.V.R. Pillay's most optimistic estimate of 50–60 mmt by the early twenty-first century may be realistic.

8.4   Conclusions

In this chapter it has been emphasized that the total weight and energy content of food harvested from the oceans and freshwater is insignificant ($<1.0\%$) when compared with that harvested from terrestrial sources. The aquatic harvest becomes considerably more important, however, if the comparison is based upon protein or, especially, animal protein. On a world-wide basis aquatic environments provide an average of 15% of the animal protein consumed by man and an even greater percentage in countries such as Japan, Norway, and Iceland which have a historical dependence upon fishery products.

Although potential food is harvested very inefficiently from both land and water, traditional methods of aquatic harvesting are 30–40 times less efficient than terrestrial farming. Modern aquaculture, on the other hand, is potentially more efficient than agriculture. Traditional fisheries techniques are essentially hunting operations with little control of either the environment or the organisms to be harvested. As a result, catches are often unpredictable and large amounts of human labour and petroleum energy must be expended. The inefficiency of fisheries harvesting might be reduced somewhat in the future by harvesting at lower trophic levels and devising new methods of concentrating widely dispersed aquatic organisms with less expenditure of energy.

The so-called 'enclosure movement' or establishment of 200-mile fishing jurisdictions should dramatically alter the pattern of fishery management in the future. Essentially, this will place 35% of the planet, formerly a common resource, under the control of single nations. Almost all stocks of fishes and shellfishes will be affected except (a) most whales, (b) 35–45% of the tuna stocks, and (c) a few other oceanic species. Initially, there may be a decline in catches within the 'enclosed' areas due to the probable exclusion of the efficient, highly technological Japanese and Russian fleets. Hopefully, the long term effect will be more careful management aimed at maintaining catches near an optimal level.

In any event, the world harvest from conventional fisheries appears to be approaching an upper asymptote of approximately 100–120 mmt. It will probably not be possible to exceed this level unless (a) new resources such as krill can feasibly be exploited, (b) new technology (i.e. fish protein concentrate) is developed and accepted, or (c) a cheap replacement is found for conventional fossil fuel energy sources.

Although the present harvest from aquaculture (6 mmt in 1975) has the potential to double by 1985, increases beyond that level are difficult to analyse and predict. It seems certain that major increases in aquaculture yields in terms of weight will require techniques which emphasize protein production of low trophic level organisms (tilapia, carp, mullet, mussels, etc.) rather than economically profitable carnivores (lobsters, pompano, trout, etc.).

In the final analysis, small to moderate increases in animal protein production from the world's conventional fisheries and aquaculture operations will probably continue through the end of the century. Any increases in the supply, however, will be more than offset by a higher demand from the rapidly growing human population.

For students interested in pursuing these topics in more detail, the following publications should provide a convenient starting point. Two useful references on the general topic of food from the seas are Rounsefell (1975) and Bell (1978). The management of conventional fisheries are described by Gulland (1971, 1974) and Cushing (1977). May *et al.* (1979) present an innovative modelling approach to the management of multispecies fisheries. Idyll (1973) details the ups and downs of the Peruvian anchovetta fishery. The potential of the Antarctic krill fishery has been analysed by El-Sayed and McWhinnie (1979). Bardach *et al.* (1972) have created an extensive work on aquaculture while Reay (1979) has provided a condensed reference on the same subject. The question of ultimate limits on fishery yields has been tackled by Ryther (1969) and Alverson *et al.* (1970). Finally, Whittaker and Likens (1975) have produced a convenient survey of total biosphere productivity and yields.

9

Strategies for Survival of Aquatic Organisms

R.N. HUGHES

9.1 Introduction

There are at least as many strategies for survival as there are species, probably more since particular species may adopt different strategies in different places or at different times. To review such a topic in one chapter necessitates a very superficial and arbitrary style. Emphasis has been placed on planktonic and benthic organisms because they reflect very clearly the opportunities and constraints associated with life in aquatic environments. Meiofauna have been omitted, in spite of their great importance in aquatic ecosystems, because less is known of their population and life-history ecology.

Virtually all attributes of an organism are of selective significance, whether body form, colour, behaviour, diet, reproductive method, habitat requirement and so on. Strategies for survival encompass all these multitudinous facets, but for convenience and generality they will be considered under the headings of feeding, reproduction and responses to stress. The headings are somewhat arbitrary and much of what is said under one will be relevant to the others.

The phrase 'strategies for survival' is used for literary expedience and of course does not imply any planning for the future on the part of an organism. So-called adaptations of organisms reflect the past influence of selection pressures. Since there is often a very good correlation between past and future events or environmental conditions, the products of past selection pressures are likely to do well in the present or near future.

A common theme of life-history strategies is the trade-off between investment of time and energy on conflicting needs. Each case is a compromise, an optimal solution for simultaneous problems. Because of the limited time or energy available to an organism, properties such as competitive ability, predator defence and productivity tend to detract from each other. Excellence in one field usually costs poorer performance in another. There is no super organism, but a whole series from the specialized expert to the jack of all trades.

9.2 Feeding

If the kinds of organisms living in aquatic and terrestrial habitats are compared, two outstanding differences become apparent. Firstly, many organisms live permanently in the water column whereas virtually no organisms live permanently suspended in air. Secondly, many aquatic organisms living fixed to the

162

substratum are animals whereas most sedentary organisms on land are plants. The first difference is due to the high density of water relative to that of air. The second difference is partly due to the higher density of water and the rapid attenuation of light passing through water.

9.2.1 PLANKTON

Prolonged passive suspension in the water column requires a small body size, generally less than the equivalent volume of a sphere 1–2 mm in diameter. Sinking rate is reduced by friction, hence smaller organisms with a greater surface area/volume ratio sink slower than larger organisms. Sinking rate may also be reduced by long spines or projections which increase the surface area/volume ratio. A slight positive sinking rate is inevitable without special buoyancy mechanisms, but turbulence in the case of phytoplankton or small zooplankton and the active swimming of larger zooplankton will override sedimentation. Passive transport in moving water masses is a prominent feature of planktonic life, bringing the advantages of vertical mixing and dispersal with the uncertainties of transportation to new areas. The rapid downstream flushing of small rivers prevents the development of planktonic communities, but water masses moving down larger, more sluggish rivers retain their integrity long enough to support plankton.

Phytoplankton

In order to photosynthesize, phytoplankton must remain within the euphotic zone. Sinking rate should therefore be kept to a minimum, not only in non-motile forms such as diatoms but also in flagellated forms where swimming to counteract sinking would consume energy that could have been used for other purposes. Flattened discs, long cylinders or filaments sink slower than spheres of similar volume (Fig. 9.1a–e). However, spheres and discs absorb more nutrient per gram per unit time than long filaments of similar diameter when sinking through turbulent water. A very slow sinking rate may favour growth since it permits more rapid uptake of nutrients as the organism passes continually through new, undepleted nutrient medium. There is consequently a trade-off between adopting forms which reduce sinking rate, thereby conserving energy or minimizing the risk of sedimentation below the euphotic zone, and forms which increase the efficiency of nutrient uptake. Buoyancy and size-specific rate of nutrient uptake both increase with smaller size, but the susceptibility to predation increases. Size and shape may therefore represent an evolved compromise to the effects of sinking rate, efficiency of nutrient uptake, predation and possibly other factors, (Hutchinson 1967).

In general, freshwater phytoplankton tend to be smaller than marine species of similar shape because their excess density over the aquatic medium is greater. In both freshwater and marine species, filamentous or discoidal shapes may reduce sinking rates, the geometry or curvature being arranged so that the organism sinks in the plane of greatest resistance. However, the effect of spines and filamentous chains is often ambiguous and may be more to deter predators rather than to reduce sinking rate (Porter 1973) (Fig. 9.1g, h). Spines made out of

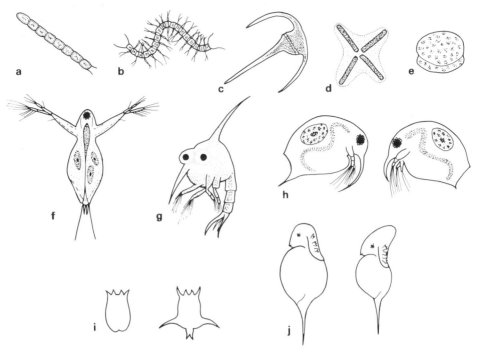

Fig. 9.1. Adaptive shapes in plankton, not drawn to scale. Filamentous chains of cells, hairs or spines reduce sinking rate in (a) diatom *Melosira*; (b) diatom *Chaetoceros*; (c) dinoflagellate *Ceratium*. Cells of the freshwater diatom *Tabellaria* (d) are connected by jelly to form a parachute-like capsule. The marine diatom *Coscinodiscus* (e) has a discoidal shape with relatively low sinking rate but efficient nutrient uptake. *Daphnia* (f) extends the antennae to reduce sinking rate. Long spines defend the crab zoea larva (g) against invertebrate predators. *Bosmina longirostris* (h) has long spines (left) as defence against predatory copepods, but shorter spines (right) where fish are the main predators. [From Kerfoot W.C. 1977 *Ecology* **58,** 303–313] Cyclomorphosis in the rotifer *Brachionus calicyflorus* (i) and the cladoceran *Daphnia retrocurva* (j). [Both after Hutchinson G.E. 1967. *A Treatise on Limnology* Vol. 2]

the siliceous frustules of diatoms or skeletal plates of dinoflagellates are probably heavy enough in the larger species actually to increase the sinking rate. Large, spiny diatoms and dinoflagellates frequently contain oil droplets and large vacuoles which increase their buoyancy. Gelatinous capsules on the other hand (Fig. 9.1d), have only a very slight excess density and therefore increase the effective surface area of the enclosed organism without significantly adding weight, so reducing the sinking rate. It is possible that gelatinous capsules also lower the rate of nutrient uptake, but such an effect has not been measured. Gelatinous capsules may be more important as an antipredator device in some cases, since encapsulated cells often pass unharmed through the guts of zoo-planktonic grazers (Porter 1973). Here the gelatinous capsules may impede the diffusion of enzymes to the cell surface.

Planktonic bacteria and blue-green algae frequently contain gas vacuoles. These are microscopic cytoplasmic vesicles which can be formed or resorbed to adjust the buoyancy of the organism. Each gas vacuole is formed by the aggregation of protein molecules to form a bi-conic capsule with a hydrophobic internal surface. Water is excluded from within the capsule so that gas diffuses inwards from the surrounding cytoplasm (Walsby 1977). Production of the proteins for vacuole formation is linked to photosynthesis. By adjusting the number of gas vacuoles, organisms can become positively, neutrally or negatively buoyant. Photosynthetic bacteria require dim light and the absence of oxygen. During the summer they become vacuolated, rising from the mud of lakes where they pass the winter, to become suspended in the dimly-lit, deoxygenated hypolimnion. Blue-green algae require oxygen and can remain suspended in the epilimnion, adjusting their depth to the optimal light intensities by controlling the number of gas vacuoles.

Zooplankton

Zooplankton feed on other planktonic organisms. Suspension among the food particles usually has to be maintained by active swimming, but the sinking rate can be reduced by extending appendages equipped with long, fine hairs. When extended, the feathery antennae of *Daphnia* (Fig. 9.1f) reduce the sinking rate by as much as 70% (Hutchinson 1967). By bending or folding at the appropriate phases of limb movement, the hairs do not impede swimming.

Phytoplankton is often patchily distributed (p. 41) so that zooplankton may have to move into new locations to secure adequate food. Certain species of marine planktonic crustaceans, ranging from copepods to decapod larvae, have been found to exhibit specific movement patterns in response to red or blue light (Dingle 1962). The red 'dance' involves relatively little activity and predominantly vertical movements, while the blue 'dance' involves greater activity and predominantly horizontal movements. Phytoplankton filters out blue light, leaving proportionately more red light, so that the red dance will bring individuals towards the phytoplankton. The blue dance will serve to disperse individuals when phytoplankton is scarce. This mechanism is only applicable to those species feeding in the euphotic zone during the day, but may be more widespread among such organisms than presently known.

Vertical migration

Vertical migration, although not always a direct response to food supplies, is certainly correlated with feeding conditions in many cases. An early idea was that by descending into deeper water layers moving at different speeds from the surface water, marine zooplankton are able to achieve considerable lateral transporation relative to their previous location at the surface. Patches of unfavourable surface water, perhaps lacking sufficient food or containing noxious phytoplankton, may be escaped by descent into deeper, faster moving water layers before the next ascent (Hardy 1962). This idea is less applicable to lakes, since the necessary deep water currents may be lacking or too weak.

Other suggestions are that vertical migration may:

1 Reduce interspecific competition by enabling individuals to select different locations that are occupied by fewer competitors or more phytoplankton.
2 Allow the interception of sedimenting particles before they are lost to the benthos.

3 Conserve energy when not feeding by allowing individuals to inhabit deeper, cooler water where metabolic costs are less and growth efficiency is higher.
4 Reduce predation pressure either by enabling prey to feed in the food rich surface waters in the dark or to elude predators by constantly changing positions.
5 Encourage matings by concentrating individuals near the surface.

Only suggestions **1** and **3** have received significant substantiation so far. Several cladocerans in oligotrophic North American lakes migrate extensively but spend most of the time in restricted sections of the water column. Heightened inter-specific competition for food in the oligotrophic lakes selects for habitat partitioning which is achieved during vertical migration. In more eutrophic lakes the same species migrate much less and mingle throughout the water column (Lane 1975). In support of the energy conservation model for vertical migration, copepods and chaetognaths have been shown to grow larger and produce more eggs in colder water. This model requires thermally stratified water and only a small energy expenditure on migration. Both requirements are likely to be met in nature (McLaren 1974). It also requires that the respiration rate decreases faster than the assimilation rate as temperature declines so that growth efficiency will increase at lower temperatures. This has yet to be demonstrated. Larvae of the midge *Chaoborus trivittatus* undergo daily vertical migrations in certain lakes but the energy conservation model was found to be inadequate in this case (Swift, 1976).

Vertical migration is not the prerogative of zooplankton, although it is far less common among plants. During calm periods when there is little vertical mixing in certain lakes, blue-green algae such as *Aphanizomenon flos-aquae* rise to the surface at night and sink during the day by means of gas vacuoles. Vertical migration may keep the blue-green algae at optimal light intensities.

9.2.2 BENTHOS

Plants

In general, angiosperms dominate the soft substrata of marginal and shallow aquatic habitats. Their supporting tissues keep emergent vegetation erect and protected from desiccation while their extensive root systems extract nutrients from the sediment and provide firm anchorage. Very unstable sediments, or those too deep for adequate light penetration, tend not to support angiosperms but are colonized by unicellular or filamentous algae whose rapid multiplication, small size and simple structure allow them to persist on shifting surfaces.

Rock surfaces on the other hand, are dominated by algae. Unicellular and filamentous algae predominate in freshwater, but the marine environment supports a diversity of macroalgae. Macroalgae are a prominent feature of temperate rocky shores where they hang in festoons at low tide, attached to the rock by a sucker-like or small root-like holdfast. Macroalgae lack the rigid supporting structures of erect angiosperms or the extensive roots of seagrasses which would outcompete them on soft substrata. But on stable rock surfaces, macroalgae can remain attached and are free from competition with angiosperms whose root systems are unsuitable for this habitat.

166

Although chlorophyll is the main pigment responsible for photosynthesis, plants usually contain accessory pigments. Of these, carotenoids and phycobilins are particularly important in brown and red algae which must photosynthesize in the filtered light penetrating seawater. Blue and blue-green light is least absorbed by water whereas the longer wave lengths of red, orange and yellow light are strongly absorbed. Carotenoids and phycobilins of complementary colour to underwater light absorb the latter more efficiently than chlorophyll, but the absorbed energy is passed on to chlorophyll for actual photosynthesis. Benthic green algae are most abundant intertidally whereas brown algae flourish well into the sublittoral, gradually being replaced by red algae with increasing depth and predominance of blue light. Although certain red algae extend to the high intertidal zone and certain green algae extend well into the sublittoral, the general sequence of green, brown to red algae has usually been regarded as chromatic adaptation to changing light quality with increasing depth. Recent work (Ramus *et al.*, 1976) suggests that this interpretation is too simple. Morphology of the seaweed appears to be at least as important as colour in determining the potential vertical distribution, whilst the actual limits to range are probably set by such factors as desiccation, grazing pressure, competition and availability of attachment sites.

Animals

Just as terrestrial and shallow water benthic plants can meet their energy requirements by remaining fixed to one spot and intercepting sunlight, so benthic animals can meet their energy requirements by remaining attached to the substratum while intercepting water-borne food particles. Such a sedentary life style is virtually impossible on land largely because the low density of air does not keep particles in suspension for very long. In water, small particles may remain suspended for extensive periods and be transported long distances. The particles are either living plankton or bacteria-coated detritus, both of which serve as food for a great variety of animals. Filter-feeders sift food particles out of suspension, while deposit-feeders gather food particles after they have settled and become incorporated into the sediment.

Filter-feeders

The array of sedentary filter-feeders is immense, with representatives from such diverse taxa as sponges, coelenterates, bryozoans, polychaetes, crustaceans, insects, gastropods, bivalves, echinoderms, tunicates (Fig. 9.2) and even vertebrates if we include such forms as *Amphioxus* or the ammocoete larva of lampreys. Food particles are intercepted as they are brought by water currents generated by tides, waves and stream flow or are filtered from currents generated by the animal itself (Jørgensen 1966). Often, filter-feeders congregate in areas where currents converge, bringing together food particles from very large catchment areas.

The majority of filter-feeders employ a mucociliary feeding mechanism whereby food particles are trapped on sticky mucus which is then transported to the mouth by tracts of cilia. Arthropods lack cilia and use meshes formed

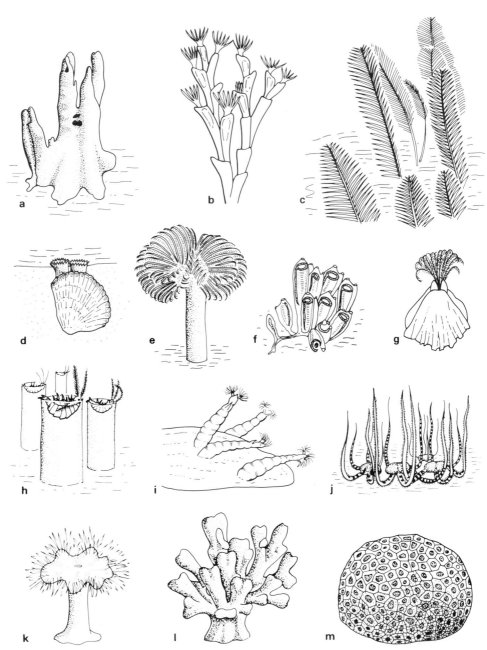

Fig. 9.2. Filter-feeders. (a) Sponge *Amphilectus*; (b) bryozoan *Bugula*; (c) hydroid *Aglaophenia* with flat surface of colony at right angles to prevailing current; (d) bivalve *Cerastoderma edule*; (e) polychaete *Bispira volutacornis*; (f) tunicate *Clavellina lepadiformis*; (g) barnacle *Balanus balanoides*; (h) amphipod *Haploops tubicola*; (i) dipteran *Simulium*; (j) brittle star *Ophiothrix fragilis*; (k) anemone *Metridium senile*; (l) multi-layered coral *Stylophora*; (m) mono-layered coral *Favia*.

168

from interlocking hairs and bristles (Fig. 9.2g, h, i). The filtration device is either external, taking the form of a more or less radially symmetrical crown of tentacles or bristles as in the coelenterates, bryozoans, polychaetes, barnacles, freshwater insects, and echinoderms, or internal, taking the form of a flat gill as in the bivalves (Fig. 9.2d) and filter-feeding gastropods or a perforated barrel-shaped chamber as in the tunicates (Fig. 9.2f). Sponges have a uniquely simple device whereby particles transported by ciliary currents stick to cells lining internal cavities. Radially symmetrical discs, shallow cones or fans formed by tentacles or intermeshing bristles, present the maximum surface area for intercepting particles from impinging currents. In many cases, the orientation of the device can be adjusted so that it always faces the changing water currents. Feather or leaf-shaped bryozoans, hydroids (Fig. 9.2c), gorgonians, pennatulids and sponges living in currents of fairly constant direction, tend to grow with their broad surface at right angles to the current so that trapping efficiency is maximized.

In clear, sun-lit tropical seas, coral tissues are colonized by symbiotic dinoflagellates which photosynthesize, deriving inorganic nutrients from the coral polyps. Coral tissues provide a steadier supply of nutrients than the open sea and are free from the very high risks of mortality associated with planktonic life. In turn, the coral polyps assimilate some of the organic matter produced by the algae. Not all corals possess dinoflagellates, but only those which do are able to secrete massive blocks of limestone and build reefs. Among the reef-building corals there is a series from those with smaller polyps which rely almost entirely on the dinoflagellates for food, to those with larger polyps which rely more on captured zooplankton for food. The plankton feeders tend to have rounded or sheet-like colonies presenting a single, continuous surface of polyps to the food-bearing water current (Fig. 9.2m). Those relying on dinoflagellates behave essentially like plants and are similarly structured to absorb light efficiently. Thus they have arborescent, multi-layered colonies, the lower branches of which absorb scattered and radiated light not striking the upper layers (Fig. 9.2l). A multi-layered colony is less suitable for filter-feeders since nearly all the impinging food particles can be intercepted by a single, continuous layer of polyps.

Deposit-feeders

Whereas filter-feeders tend to live on hard substrata or in sediments where there is little silt to clog the delicate filtering mechanisms, deposit-feeders are able to cope with silty substrata which they ingest to extract the microorganisms coating or otherwise associated with the detritus or mineral particles. Most of the ingested sediment is indigestible so that copious amounts must be processed. Sediment may simply be swallowed as in many polychaetes and freshwater oligochaetes, transported to the mouth on mucus-coated appendages as in holothurians, burrowing sea-urchins (Fig. 9.3e) and cirratulid polychaetes (Fig. 9.3a), or swept up with bristly appendages as in certain amphipods, isopods, decapods and chironomid larvae (Fig. 9.3b, d). Tellinid bivalves have long extensible siphons for sucking up the fine surface layer of sediment which is then processed on gills able to handle large quantities of silt (Fig. 9.3c). Because deposit-feeders

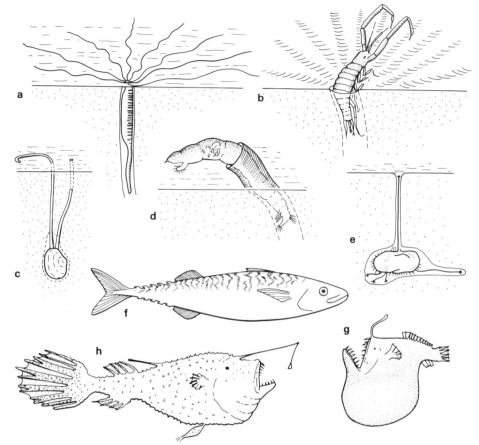

Fig. 9.3. Deposit-feeders (a) Polychaete *Amphitrite*; (b) amphipod *Corophium volutator*; (c) bivalve *Scrobicularia plana*; (d) dipteran larva *Chironomus*; (e) sea urchin *Echinocardium cordatum*. Predators: (f) Mackerel (pursuer); (g) deep-sea angler (ambusher) *Melanocetus johnsoni* with stomach distended by large prey item; (h) deep-sea angler *Ceratias holbolli* with tiny parasitic male underneath.

need to process such large amounts of sediment and also tend to live in unstable substrata, they are generally more mobile than filter-feeders, showing greater morphological adaptations for locomotion, lacking radial symmetry and never forming true colonies.

Predators

Herbivory and carnivory are essentially similar, both denoting the feeding of an animal on other living organisms and in this sense may be regarded as kinds of predation. Predators may consume entire prey organisms as in zooplankton eating unicellular algae or fishes eating zooplankton, or they may consume only part of the prey organism without killing it, as in grazers of sedentary colonial animals and benthic algae.

Both grazers and non-grazers show a range of feeding strategies from extreme dietary specialization to extreme generalism. Dietary specialization is more

170

appropriate where specific prey are predictably abundant, while generalization is more appropriate where prey abundance is low or unpredictable. For a given predator, the array of potential prey species can be ranked in decreasing order of dietary value, i.e. energy yield per unit handling time. When the highest ranking prey are abundant, the predator may specialize on them and ignore lower ranking prey; but if the preferred prey become scarce, the diet will expand to include less preferred prey so that the predator becomes a more generalized feeder. The Pacific N. American starfish *Pisaster ochraceus* prefers mussels but will include barnacles and gastropods sequentially in its diet as the preferred prey become scarce.

If alternative prey are not too dissimilar in dietary value, but one type is much more abundant than the other at a given time, then the predator may specialize on the abundant prey while ignoring the scarce prey. If the relative abundances of the prey are reversed, the predator may switch to specialize on the second prey type. The dogwhelk *Acanthina spirata* switched its diet from mussels to barnacles and vice versa as relative abundances of the prey were changed in the laboratory (Murdoch 1969). Switching is a potential stabilizing mechanism since the predator will cease to feed on prey which become too scarce as long as alternative prey are available.

Predators tend to be either pursuers or ambushers. In general, pursuers, which spend considerable energy chasing prey, may be expected to specialize on prey with high dietary values, perhaps becoming morphologically adapted to capturing particular prey species very efficiently. Ambushers, which spend little energy capturing prey but cannot select which prey shall pass by, may be expected to generalize and accept all capturable prey. Mackerel (Fig. 9.3f) and tuna are good pursuers, specializing on prey fishes of the appropriate size. Deep-sea predatory fish such as gulpers, swallowers and anglers are ambushers par excellence. At these great depths prey are very scarce indeed, so that pursuit and selective feeding would be uneconomical. The predators have become suspended traps, most of the body muscles and skeleton being atrophied except for the jaw apparatus. The mouth has such an enormous gape and the stomach is so distensible, that prey considerably larger than the predator can be swallowed (Fig. 9.3g) (Marshall 1954).

9.3   Reproduction

9.3.1   ASEXUAL VERSUS SEXUAL REPRODUCTION

Reproduction in plants and animals may be asexual, where a single parent produces offspring genetically identical to itself, or sexual where gametes from two parents combine to produce genetically unique progeny. In general, asexual reproduction is used for rapid population growth. There are no males unable to give birth so that every individual can contribute to population growth. The offspring may separate from the parent, as when clones of diatoms or cladocerans expand rapidly in the competitive race to exploit resources while favourable conditions last, or they may remain attached to the parent as in macroalgae or corals, forming progressively larger colonies which are better able to compete for resources, resist predation or withstand physical stresses.

Sexual reproduction requires the expenditure of energy on gonads or ancillary sexual organs and requires time for courtship, copulation, fertilization and embryological development. Sexual reproduction therefore results in a lower potential rate of increase than asexual reproduction. On the other hand, because of recombination at meiosis, sexual reproduction results in genetic variation among the offspring. This genetic variation may be of considerable advantage to organisms which undergo wide dispersal at some stage in their life history. In such cases, the disseminules may settle unpredictably among a wide variety of microhabitats and genetic variability among them will maximize the chances that at least some will be well matched to the conditions at the settlement sites. Similarly, if environmental conditions change unpredictably from season to season, then future generations of ephemeral or annual organisms may grow up in environmental conditions different from those experienced by the parental generation, so that genetic variability among the offspring will again be advantageous. It is not surprising, therefore, that sexually produced larvae are used for dispersal in many sedentary animals and that annual or ephemeral species often produce sexual spore-like stages or resistant zygotes which remain dormant, typically undergoing passive dispersal, during unfavourable periods or during the non-growing season. Small, undifferentiated embryos may conveniently be covered with a tough outer coating to form 'spores' resistant to desiccation, extreme temperatures, or prolonged food shortage. They are usually produced when the environment begins to deteriorate, heralding the onset of prolonged harsh conditions such as drought or cold. During favourable periods it is better for the annual or ephemeral species to grow asexually, thereby maximizing clonal size. Sexual reproduction during the favourable periods would detract energy from asexual growth so that the clone would reach a smaller final size. Sexual fecundity increases with clonal size and is therefore maximized by deferring gamete formation until rapid asexual growth is impeded by deteriorations of the physical environment or by intense intra- and interclonal competition for food.

Freshwater zooplankton and hydras produce resistant sexual eggs when ponds begin to dry up or when food supplies and temperatures drop at the onset of winter. The resistant bodies remain dormant until favourable conditions return. Freshwater sponges and some intertidal or shallow sublittoral marine sponges, hydroids, bryozoans and tunicates also produce resistant spore-like bodies, buds and stolons at the onset of winter, but these are usually produced asexually. In such cases, the asexual overwintering bodies are really a dormant phase of the colony which regenerates from them when favourable conditions return. The regression of the colony into overwintering structures is analogous to perennial herbs overwintering as bulbs, corms or rhizomes. Although perennial benthic colonies may achieve some measure of dispersal by passive transportation of dislodged asexual overwintering structures, the latter tend to remain attached to the substratum and serve to recolonize quickly the space previously occupied by the fully developed colony. Such pre-emption of space is of great importance in competition with other sedentary organisms. Dispersal in these animals is mainly via larvae produced sexually by the fully developed colony.

In summary, sexual reproduction in annual or ephemeral clones results in the production of resistant zygotes or spore-like structures in the advent of hostile conditions, whereas sexual reproduction in perennial colonies results in the production of motile larvae during favourable periods. In both cases, it is the main disseminules that are produced sexually and are therefore endowed with the genetic variability which is advantageous in colonizing microhabitats of unpredictable quality.

A partial exception to the above scheme is seen in freshwater bryozoans and perhaps some freshwater sponges, where there is a second major phase of dispersal in addition to that achieved by the larvae. The second dispersal phase involves spore-like bodies produced asexually in the advent of harsh conditions. Although some of these 'spores' may remain attached to their original site and re-establish the functional colony when conditions ameliorate, others may become detached and carried by water currents, wind or mobile animals to new areas. Certain freshwater bryozoans are known to produce two types of 'spore' (statoblasts), those which remain attached to the original site and others which float by virtue of air spaces in the outer coating. Drifts of statoblasts up to 1 m wide, and extending for as much as 800 m have been seen on the shores of Lake Michigan. Possibly the sexually produced larvae serve for relatively short-distance dispersal while the more resistant 'spores' serve for hazardous long-distance dispersal. Unlike larvae, resistant 'spores' may survive overland dispersal by wind or animal transportation to other bodies of water. Why encapsulated zygotes, which would have genetic variability, are not produced for long-distance disperal by these animals remains a mystery.

Clones and modular colonies

Whether asexually-produced offspring remain attached to the parent and form a colony of modular construction, or whether they detach from the parent to form a clone of free individuals depends on the life-style of the organism. Colonies of modular construction, such as colonial hydroids, bryozoans, tunicates, macroalgae and higher plants, are able to compete effectively for space, suspended food particles, or light by lateral sheet-like growth over the substratum or by vertical, arborescent growth overtopping neighbouring competitors. Continually expanding modular colonies, by increasing the surface area of filtering zooids or photosynthetic tissue (Fig. 9.4a, b), offer a very efficient way in which sedentary organisms can compete for resources. Moreover, because all modules are genetically identical, some of them can loose their reproductive capabilities and become specialized entirely for feeding or defence without violating the laws of individual, as opposed to group selection (Fig. 9.4c).

Modular colonies are less suitable for mobile animals and are confined to a few aberrant benthic bryozoans which are able to creep slowly over the substratum and to a minority of pelagic filter-feeders and carnivorous coelenterates which drift more or less passively. Modular colonies are more frequent among phytoplankton which lack the complex appendages of animals, but colony size must be restricted to avoid excessive sinking rates and mechanical fragility. Locomotion imposes severe mechanical constraints on the form and size of the body, so that most mobile animals reproducing asexually release their offspring as replicas of themselves, each well designed for swimming or crawling.

Fig. 9.4. Modular colonies and clones. (a) Brown alga *Fucus spiralis* with repeated bifurcations tipped with reproductive organs; (b) hydroid *Bougainvillea* with repeated zooids all of similar kind, the older stems of the colony are fouled by diatoms; (c) hydroid *Hydractinia echinata* with specialized zooids for feeding (long tentacles), defence (cluster of knob-like tentacles) and reproduction (large vesicles, female at left and male at right of colony); (d) *Hydra* clone spreading by asexual budding along lily stalk; (e) *Daphnia* clone produced parthenogenetically.

Clones of separate individuals (Fig. 9.4d, e) have a dispersal advantage and, with their high multiplication rate, are of selective value when exploiting temporary or seasonal resources. Members of a growing clone may experience intraclonal competition which retards multiplication rate or, in very restricted environments, may even prevent further clonal expansion. A classical example is given by the logistic, asymptotic growth curves of the *Paramecium* clones reared in culture vessels by Gause. Intraclonal competition is often confused with intraspecific competition. However, intraspecific competition implies competition between different genetic individuals. Intraclonal competition adjusts the size of the clone, i.e. of a single genetic individual, to the supply of resources and is a growth regulating mechanism analogous in some respects to those regulating body size. It is therefore not deleterious to the genetic individual in the sense of intraspecific competition.

Planktonic larvae

Dispersal enables sedentary organisms to exploit newly available substrata in different localities and insures against deterioration of the parents' patch of habitat. Zygotes are convenient starting points for the embryological development of larvae which, being evolved in the plankton, are morphologically quite different from the parents (Fig. 9.5a–c). Dispersal by sexually produced planktonic larvae is therefore a common trait among marine benthic invertebrates.

Marine planktonic larvae fall into two categories: planktotrophic larvae which remain in the plankton for long periods of up to several months during which they feed on other planktonic organisms, and lecithotrophic larvae which stay in the plankton for short periods of a few hours, relying on stored energy rather than feeding on plankton (Thorson 1946). Planktotrophic larvae are suitably designed for very long distance dispersal in offshore currents. They may have to wait some time before they escape local, reversible tidal currents and before they are brought into contact with substrata suitable for settlement. Consequently they must feed on plankton to meet their maintenance energy requirements. A plausible idea is that planktotrophic larvae are able to take advantage of phytoplankton blooms, thereby increasing growth rate and survivorship relative to non-feeding larvae, but this remains to be substantiated. Lecithotrophic larvae are appropriately designed for shorter distance dispersal by tidal currents. Because tidal currents are reversible, the average distance dispersed from the point of release increases rapidly only for the first 6 hours where there are semi-diurnal tides or 12 hours where tides are diurnal, after which further increases are small. Consequently there is no point in such larvae remaining in the plankton for more than a few hours. Further delay would gain little extra dispersal but would consume energy and prolong exposure to predation. Intermediate strategies between long and short duration in the plankton are therefore rare (Crisp 1976).

The probability of being eaten multiplies as the duration of the planktonic phase increases. However, planktotrophic larvae are produced in much larger quantities than lecithotrophic larvae. This need not cost more energy since planktotrophic larvae are endowed with much smaller energy reserves than lecithotrophic larvae, but adequate data are lacking on this point.

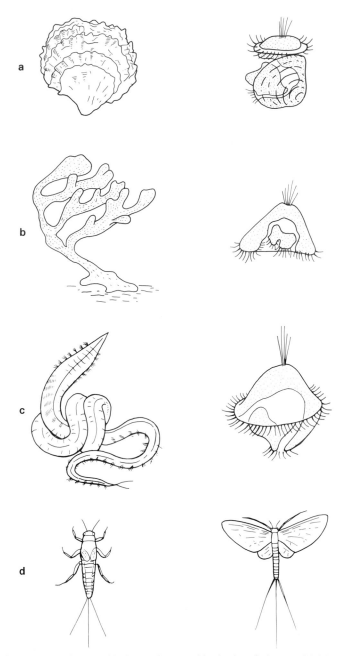

Fig. 9.5. Dispersal. Marine benthic invertebrates with planktonic larvae; (a) bivalve; (b) bryozoan; (c) polychaete. The larvae are about 0.5–1.5 mm overall diameter. Freshwater insect (mayfly) with dispersing adults: (d) adult and larva are of similar size.

In addition to predation, planktonic larvae also face the risk of missing suitable settlement sites. Detection of a suitable settlement site is enhanced in many species by a chemoreceptive response to the presence of adults as in barnacles and oysters, or to a specific substratum as when the serpulid worm *Spirorbis borealis* settles on the brown alga *Fucus serratus*. Settling near adults, apart from running the risks of intraspecific competition or being eaten by the adults, ensures that the larvae are colonizing a patch of habitat proven for its overall suitability and causes conspecifics to aggregate, thus facilitating fertilization. Larvae usually require some minimum time in the plankton before they will settle, even in the presence of suitable substrata, thus ensuring dispersal. Thereafter, settlement will usually occur at the earliest opportunity but can be delayed for a while if suitable sites are not encountered. Site specificity tends to decrease with increased duration of site deprivation but the chances of survival or good growth will decrease in suboptimal sites.

The advantages of dispersal increase with the spatial discontinuity or temporal unpredictability of habitat patches. However, dispersal may not be the only factor selecting for planktonic stages. The planktonic larvae of many demersal fish (Fig. 9.6b) are able to find an abundance of suitably small food particles in the plankton and are free from competition or predation by older stages. The larvae suffer the characteristically high planktonic mortality rates but are produced in copious quantities.

Dispersal via planktonic larvae is rare in freshwater. Lakes lack the strong currents necessary for long distance dispersal by passive transportation. Moreover, similar kinds of habitat vary much less in quality from place to place in lakes than in the sea. The unidirectional flow of rivers makes planktonic dispersal disadvantageous. Unionid freshwater mussels provide an interesting exception where larvae are puffed out by the parent on to passing fish, to which they become attached and are carried until ready to drop to the river bed as newly metamorphosed mussels. Ponds are too small for planktonic dispersal to be advantageous and are isolated by dry land which can only be crossed by impervious spores, cysts or eggs blown by wind or carried by attachment to larger animals.

In contrast to marine benthic invertebrates, the roles of larva and adult are reversed in freshwater aquatic insects. The majority of aquatic insects spend most of their life as benthic larvae which are the main trophic stages. The brief, winged adult stage undergoes sexual reproduction and aerial dispersal (Fig. 9.5d).

9.3.2 PARENTAL INVESTMENT PER OFFSPRING

The amount of energy invested per offspring by the parent decreases in the order of brooders (Fig. 9.6a, c), oviparity with direct development (Fig. 9.6d), oviparity with lecithotrophic planktonic larvae, oviparity with planktotrophic larvae (Fig. 9.6e).

Species living in more uniform or extensive habitats which do not tend to deteriorate unpredictably, may gain less from dispersal and more from propagules which can compete effectively for resources, resist predation or withstand physical stresses. Such species lack the risky planktonic stage and produce fewer, larger eggs well endowed with energy reserves, which hatch into

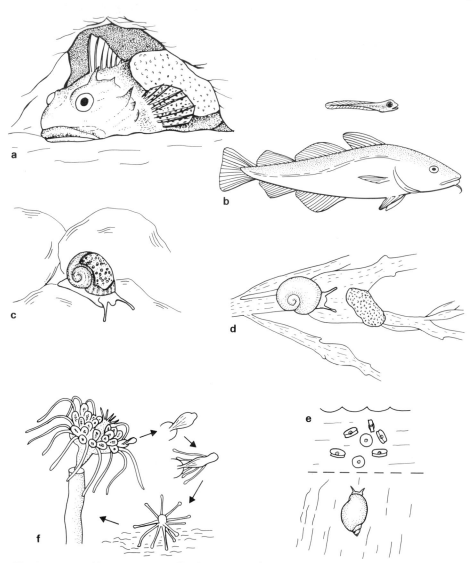

Fig. 9.6. Parental investment per offspring. (a) Male sea-scorpion (sculpin) *Cottus bubalis* guarding eggs; (b) cod with abandoned planktonic larvae; (c) winkle *Littorina rudis* (= *saxatilis*) with part of shell removed to show brooded young; (d) *Littorina obtusata* (= *littoralis*) with egg mass attached to alga *Ascophyllum nodosum*; (e) *Littorina neritoides* with broadcast planktonic eggs; (f) hydroid *Tubularia* with benthic larva.

relatively large benthic larvae, e.g. the actinula larva of the hydroid *Tubularia* (Fig. 9.6f), or develop directly into juveniles essentially similar in form to the adults as in the snail *Littorina obtusata* (Fig. 9.6d). The risk of juvenile mortality and susceptibility to competition are reduced even further by brooding the young to an advanced stage of development, e.g. *Littorina rudis* (Fig. 9.6c).

Although parental investment per offspring may often be correlated with habitat stability, degree of competition, and harshness of the physical environment, other factors will also influence the degree of investment. High parental investment by building nests, brooding young or guarding eggs is common among fish with restricted home ranges in structurally complex environments, e.g. crevice-dwelling blennies, sculpins (Fig. 9.6a) or the numerous territorial coral-reef fishes. Low parental investment with the spawning of vast quantities of unprotected eggs is more characteristic of nomadic fish (Fig. 9.6b) which do not become familiar with local habitat details nor live in places affording an abundance of protected nest sites. Low parental investment per offspring among nomadic fish is also correlated with their frequent adoption of planktonic life in the larval stages.

9.3.3 SEMELPARITY VERSUS ITEROPARITY

Organisms may reproduce sexually once and then die (semelparity) as with annuals, or reproduce more than once (iteroparity) as with perennials. The appropriate method is largely determined by the ratio of juvenile to adult survivorship. Semelparity is more appropriate where the ratio is high, denoting relatively low adult survivorship, whereas iteroparity is more appropriate when juvenile survivorship is relatively low (Stearns 1976). Semelparity relies on the survival of the young to maturity whereas iteroparity relies on survival of the adults, which is achieved by investing proportionately more energy on the adult body and less on reproduction. In many freshwater insects such as dipterans, caddisflies, stoneflies and mayflies, the dispersing adults are more at risk to mortality than the benthic larvae and semelparity is the rule. The benthic adults of marine sedentary invertebrates, on the other hand, are less at risk to mortality than the dispersing larvae and iteroparity predominates among these organisms.

9.3.4 MATING PROBLEMS AND HERMAPHRODITES

Sedentariness or low population density will reduce the chances of fertilization between two individuals. In several deep-sea angler fishes, e.g. *Ceratias holbolli* (Fig. 9.3h), which live at exceedingly low population densities, the dwarf male becomes organically attached to the female, deriving nourishment from her and remaining with her throughout reproductive life. The rate of encounter between the sexes is so low that once contact has been made it is advantageous to maintain it.

Simultaneous hermaphroditism, by doubling the chances of meeting another individual of opposite sex, is another way of increasing the probability of cross fertilization and is found among most sedentary animals and in very sparsely distributed motile species. Whereas many organisms are simultaneous hermaphrodites with male and female gonads functional at the same time, others spend the first part of their life as one sex and later change to the other. Those starting as females are called protogynous sequential hermaphrodites, while those starting as males are called protandrous sequential hermaphrodites. Sex reversal may occur because younger individuals function better as one sex and older individuals function better as the other sex. Most sequentially hermaphrodite invertebrates are protandrous, e.g. *Hydra*, the common limpet *Patella vulgata* and the slipper limpet *Crepidula fornicata*. It is often assumed that because sperm are so tiny, sufficient quantities can be produced by young, small

179

animals while at the same time leaving an adequate energy surplus for somatic growth. On the other hand, intermale competition would tend to favour older, larger, more fecund males which did not change sex. The phenomenon is little understood, but probably involves an advantage gained by increasing the likelihood of encounters between opposite sexes.

Quite different selection pressures account for protogynous sequential hermaphroditism in wrasses and parrot fish of tropical reefs. In these fish, only territory-holding males are accepted as mates by the females. Male reproductive success therefore depends on being able to defend a territory from competing males and this can only be achieved by large, strong fish. Young, small fish could not compete for territories, so they function as females until the critical age or size is reached (Warner 1975).

9.4 Responses to stress

9.4.1 PHYSICAL FACTORS

Physical stresses in aquatic environments usually amount to fluctuations and extremes of salinity or temperature, lack of oxygen, desiccation, silting, dislodgment by waves or currents. Anatomical, physiological and behavioural responses to physical stresses are diverse and beyond the scope of this chapter. But as well as causing such responses, physical stresses often have important effects on life histories.

Predictable, seasonal stresses such as low winter temperatures may select for annual life histories whereas more unpredictable periods of stress, such as the shrinkage of ponds during droughts, tend to select for shorter lived ephemeral species. Both annuals and ephemerals usually have resistant spore-like bodies or eggs which remain dormant throughout the stressful period. Examples are found among freshwater algae, hydras, bryozoans, sponges, cladocerans, copepods, brine shrimps and among intertidal or shallow sublittoral marine algae, hydroids, bryozoans, sponges, tunicates and rock-pool copepods. Since populations are decimated during the stressful periods, they recurrently enter a phase of expansion when favourable conditions return. Under these conditions, individuals with higher reproductive rates will be fitter than those with lower rates of increase. Selection maximizes both rate of growth to maturity and fecundity, i.e. productivity, and this is often achieved by rapid growth by budding as in coelenterates, bryozoans, tunicates and other benthic invertebrates, or by parthenogenesis as in cladocerans and copepods.

9.4.2 COMPETITION

Whereas expanding populations in recurrently emptied patches of habitat gain selective advantage by maximizing productivity, equilibrial populations in more stable habitats gain competitive ability by aggressively defending resources, maximizing efficiency of resource usage, or ousting neighbours by more rapid or taller growth. Such strategies cost energy which is then denied to reproduction, so that there is a trade-off in stable environments between competitive ability and productivity.

Aggressive defence of resources is superbly illustrated by the South African

180

limpet *Patella longicosta* which feeds on the brown alga *Ralfsia* growing within its territory. If other foraging limpets of the same or different species wander into its territory, *P. longicosta* advances on them, slowly pushing them away (Branch 1975). Such aggressive behaviour defends the resource but costs time and energy which could otherwise be spent on feeding and production. The magnitudes of such costs have never been measured, but are perhaps quite small in the case of *P. longicosta*.

Efficiency of resource usage is usually achieved by increasing specialization on a narrower part of the resource spectrum. Deposit-feeding snails of the genus *Hydrobia* choose food particle sizes roughly in proportion to their body size. When occupying separate habitats *H. ventrosa* and *H. ulvae* reach the same average body size and feed on similar particle sizes; but when coexisting on the same mud flat, the average size of *H. ventrosa* becomes smaller and that of *H. ulvae* becomes larger, reflecting an increased specialization on smaller and larger food particle sizes respectively (Fenchel 1975). Such genetically-based 'character displacement' reduces overlap in resource usage, alleviating interspecific competition and allowing coexistence. Intraspecific competition causes the relaxation of character displacement in isolated populations of each species.

The successful replacement of fast-growing, rapidly colonizing but small and flimsy green algae on rocky intertidal shores by slower growing, larger brown algae illustrates the competitive value of supporting structures such as stipes and midribs which allow growth to a larger size and the shading of competitors. The cost of structural tissues are lower productivity and slower colonization so that the large brown algae tend to perform less well than green algae in very unstable or disturbed patches of habitat.

Persistent physical stress, rather than selecting for fugitive species with high productivity, will select for stress-resisting mechanisms which usurp or forfeit energy, so reducing productivity. An example is the high intertidal barnacle *Chthamalus stellatus* which resists desiccation partly by investing in robust, tightly fitting skeletal plates. *Balanus balanoides* has been less substantial, more loosely fitting plates which would allow too great a water loss at high shore levels, but which apparently facilitate faster growth at lower levels. At these low levels it grows faster than *Chthamalus* and outcompetes it for space. *Chthamalus*, however, is able to occupy the high shore refuge free from interspecific competition.

Community succession illustrates the series that exists between the opportunistic colonizers of empty habitats on the one hand and the superior competitors in fully occupied, undisturbed habitats on the other. The early colonists are gradually replaced by superior competitors. But even among the latter, there is usually a hierarchy of competitive ability headed by one or two potential competitive dominants. If habitat patches are left undisturbed, the potential dominants will eventually monopolize the resources, excluding all other competitors. By eliminating local patches of potential dominants, accidents of physical or biological origin make resources available for competitively inferior species. For example, wave action and predation by starfish repeatedly clear away patches of mussels on the mid shore of the Pacific North American coasts and on the low shore of certain British coasts. Certain kelps on the low shore of Pacific North American coasts are kept from complete dominance by grazing sea-urchins, while total monopoly of space by a competitively dominant sponge in the Antarctic sublittoral is prevented by grazing starfish and nudibranchs (Dayton 1975). In each case, competitively inferior species are able to coexist with potential dominants due to the repeated interruption of succession. Of course, disturbances which are too frequent or too extensive will be detrimental to all species and will reduce, rather than increase, the number of species present. Shifting sand bars, for instance, are very depauperate in species.

9.4.3 PREDATION

Antipredation mechanisms are employed by virtually all organisms, at least during the more vulnerable stages of their life history. The diversity of mechanisms is bewildering, including features such as spines, heavy shells, repellent taste or texture, Mullerian and Batesian mimicry, crypsis, behavioural responses and so on. All these mechanisms require an investment of time or energy which is denied other functions such as competition, growth or reproduction. The degree to which antipredator mechanisms are developed will therefore represent an optimal solution to several conflicting demands on time or energy. For example the cladoceran *Bosmina longirostris* has two forms in Lake Washington (Fig. 9.1h) (Kerfoot 1977). The long-spined form lives over deep open water where predatory copepods are abundant. The long spines interfere with the intricate feeding appendages of the copepods. The short-spined form lives in shallow water among weed beds where fish reduce the densities of all zooplankton, so that predation on *Bosmina* by copepods is slight. The longer spines of the open water form require extra yolk and larger eggs for their development. Because of limited space in the brood chamber, the long spined form with its larger eggs has smaller broods and hence a lower potential rate of increase than the short spined form. Consequently the short spined form outcompetes the long spined form in shallow water where long spines confer no significant protection against fish predators.

Protection from small invertebrate predators can be achieved by developing spines as in *Bosmina* or by becoming too large to eat. Neither of these mechanisms, however, is likely to be effective against large predators such as fish. Vertebrates will tend to prefer larger, energetically more rewarding zooplankton as food and, being mainly visual predators, will cause selection for smaller zooplankton and cryptic properties such as transparency. The size-frequency structure of zooplankton in certain North American lakes has been shown to depend on the kind of predators present. Lakes where fish or amphibians are abundant lack the larger species of zooplankton, whereas those lakes where vertebrate predators are few or absent are populated by larger zooplankton with some measure of resistance to invertebrate predators (Dodson 1974a).

Certain marine benthic invertebrates are able to gain a size refuge from even quite large predators. The barnacle *Balanus cariosus* and the mussel *Mytilus californianus* are preyed upon by dogwhelks, *Thais* spp., but the shells of those prey which survive long enough become too thick for dogwhelks to penetrate. Size refuges from predation may be quite important in preventing local extinctions of prey.

182

Many freshwater planktonic organisms including certain algae, rotifers and crustaceans, alter their body form between successive generations during the year. This phenomenon, known as cyclomorphosis, involves the development or exaggeration of spines and prominences in certain generations and their regression in others (Hutchinson 1967). An increasing body of evidence (Dodson 1974b) suggests that cyclomorphosis is often a response to predation pressure. Invertebrate predators usually have a very narrow range of preferred food sizes and, owing to the precise structure of the prey-catching apparatus, the shape of the food item may be quite critical. The development of spines or protruberances will therefore jam the predator's feeding mechanism. The predatory rotifer *Asplanchna* feeds on a smaller rotifer *Brachionus*. When *Asplanchna* becomes abundant, *Brachionus* is able to detect waste metabolites produced by the predator and responds by growing long anterior and posterior spines (Fig. 9.li).

The preferred prey size of invertebrate zooplanktonic predators reaches only about 1.5–2 mm, whereas limitations of the vertebrate eye probably preclude intense predation by vertebrates on zooplankton smaller than about 1 mm. Cladocerans with a body size of 1–2 mm are on the border line between immunity to invertebrate predation and high susceptibility to vertebrate predation. Consequently, some species of *Daphnia* and other cladocerans decrease the core body size, while increasing the development of spines and protruberances. The visible body core thus becomes too small for vertebrate predators, while the less visible spines and hyaline protruberances (Fig. 9.1j) maintain an overall size too large for mechanically-feeding invertebrate predators (Dodson 1974b).

The cyclomorphic development of spines etc. thus corresponds to periods when predators become abundant. The ornamentation regresses when predators become scarce, presumably because it is disadvantageous when not needed for defence. The precise nature of this trade-off has not yet been worked out, but possibly involves the energetic cost of developing spines, increased drag while swimming, or increased sinking rate.

9.4.4 FOULING

Sedentary aquatic macroorganisms themselves provide surfaces attractive to settling spores and larvae (Fig. 9.4b). This is seen on the low intertidal alga *Fucus serratus* which supports a characteristic fauna of serpulid worms, bryozoans, hydroids, tunicates and sponges, or on the flat fronds of the sublittoral bryozoan *Flustra foliacea* which also supports a characteristic assemblage of sedentary species. Increasing loads of epifaunal organisms will interfere with light absorption or filter-feeding and, by causing drag, will increase the risk of dislodgement by currents or wave action. Many sedentary organisms possess antifouling mechanisms such as mucus-coated surfaces, repellent chemicals, spines, the pincer-like zooids of bryozoans or pedicellariae of echinoderms, and the stinging nematocysts of coelenterates. The chances of fouling will also be reduced by the mere filter-feeding activities of the potential hosts during which settling spores and larvae will be consumed. Because of the ever-present risk of fouling, antifouling mechanisms may be expected to be widespread among sedentary organisms, especially among the longer-lived species. It is known that the fronds of certain macroalgae contain antibacterial substances and that a proportion of individuals among populations of *Fucus serratus* is often much less heavily fouled than the rest. But our knowledge of antifouling mechanisms is restricted to a few such isolated observations and the topic remains a fascinating challenge to future researchers.

This chapter has touched lightly on just a few topics within the enormous domain of 'strategies for survival'. Exciting discoveries and ideas are being made in many other relevant fields. To experience this excitement requires an even broader approach than that adopted here. Some stimulating references within the general area of evolutionary ecology include Harper (1978), Krebs and Davies (1978), Maynard Smith (1978) and Pianka (1978).

10

The Total Aquatic System

K.H. MANN

10.1 Energy and nutrients

When energy flow is mentioned in ecology, we usually refer to the fixation of solar energy in photosynthesis and its transfer through food webs. There is, however, a much greater flow of energy in physical processes which is absolutely critical to the functioning of aquatic ecosystems. Stated briefly it is the energy required to carry plant nutrients from deeper waters, where they tend to accumulate, to surface waters where they can be used in photosynthesis. In a conceptual model that represents biological production as a flux of solar energy, the gate or valve that controls the flow of energy is operated by the flux of plant nutrients. In almost all situations, plant productivity in aquatic ecosystems is limited by the supply of plant nutrients for some part of the year. The greater the depth of the water body, the greater the energy required to bring up the nutrients. This is one reason why shallow bodies of water tend to be more productive.

Energy for upwelling is obtained primarily from four mechanisms: convective cooling, wind-induced currents, river runoff, and tidal currents. The first three are driven by solar energy, the last by the gravitational pull of the moon. We can think of them as physical energy subsidies which assist the biological processes.

Fig. 10.1 illustrates these mechanisms for bringing nutrient-rich deep water to the surface. The first is convective cooling, which can only take place if there has been equivalent warming at some other time or place. Hence the driving forces are seasonal or spatial differences in solar radiation. Atmospheric cooling leads to increasing density in the upper layers, and a tendency for surface waters to sink. There has to be a compensatory rise of water from lower levels. This mechanism operates in the autumn in temperate latitudes, bringing about an annual renewal of the nutrients of surface waters.

The second mechanism is wind-induced mixing. Wind over water causes a surface current in the same direction. Friction between the moving water and the stationary lower layer generates vertical turbulent mixing, bringing some of the deeper water to the surface. If the wind-driven current moves away from a shore line, there will be a strong upwelling of deep water to replace that leaving the shore. Wind-induced mixing and upwelling complement convective cooling as a way of mixing temperate lakes after summer stratification. Arctic lakes may mix freely throughout the ice-free period, while tropical lakes may be per-

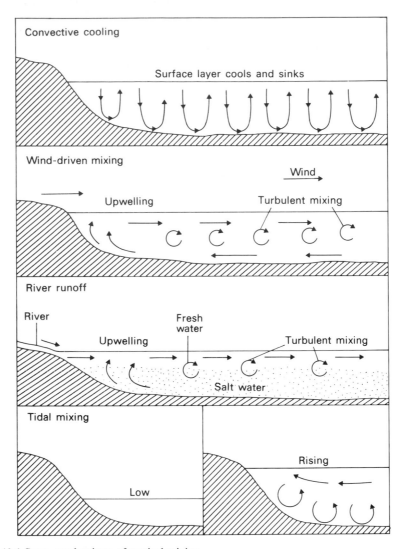

Fig. 10.1 Some mechanisms of vertical mixing.

manently stratified. Analogous periods of stratification and mixing are found in coastal waters.

On a global scale, wind-driven currents cause huge circulation patterns in the world's oceans (Fig. 7.1) with clockwise circulation in the northern hemisphere and anticlockwise circulation in the southern hemisphere. In places where the main currents move away from land masses (primarily on the eastern sides of the basins) there are major areas of upwelling.

The third mechanism is river runoff. At places where rivers discharge to the sea, a layer of fresh water moves across the more dense salt water. As with wind-driven currents, the shear between the two layers brings about vertical turbulent mixing. In estuaries of the appropriate shape, there is a steady shoreward movement of deeper water to compensate for the outward movement at the surface. The net result is a very considerable upwelling of nutrients, which partly accounts for the high productivity found in many estuaries.

Finally, in shallow bays and estuaries where the rise and fall of the tide is considerable in relation to the volume of the basin, tidal currents may be sufficient to keep the waters well mixed at all times. Nutrients regenerated from the benthos are then continually made available to primary producers.

There are terrestrial analogues of this indirect effect of physical energy on biological production. In a mature forest with a dense canopy many metres above the ground, there is the same spatial separation of photosynthesis at the top and nutrient regeneration in the litter layer. The gap is closed by raising the nutrients in the transpiration stream within the trunks of the trees, and the energy from this comes largely from sun and wind. We are familiar with the idea that in a less natural situation primary production on a farm can be greatly enhanced by the addition of fertilizers. Production and distribution of fertilizers requires the expenditure of large amounts of fossil fuel energy. Man is carrying out, at considerable cost, the same process of nutrient transport that in aquatic systems is done for nothing by natural forces.

10.2   Grazing and detritus food webs: are they distinct?

Macrophytes are a distinctive component of the fringing communities of aquatic ecosystems (Chapter 4). Many of these aquatic macrophytes have evolved from land plants and have the characteristic structural tissues needed for support in air, together with conducting tissue for movement of the transpiration stream. Such tissues contain considerable amounts of rather indigestible substances, e.g. cellulose and lignin. Macrophytes also tend to evolve the capacity to synthesize substances which are unpalatable to animals, giving them a chemical defence against herbivores. Hence, on land and in the fringing communities of aquatic systems, we find that the majority of large plants are not extensively grazed by animals. Instead they die, producing detritus which is first attacked by bacteria and fungi, and only secondarily attacked by invertebrates (Chapter 4). Some of the latter specialize in shredding the detritus which increases the surface area available for attack by microorganisms. Others ingest particles of detritus and retain the microorganisms, passing the indigestible plant material out in their faeces. The whole process leads to the eventual production of very small detritus particles which may be consumed by filter feeders either in the plankton or in the benthos.

It is interesting to compare this process with a grazing food chain on land. For example, grass is ingested by a ruminant, and the ruminant has to cope with the same problem of the indigestibility of the structural tissues. It over-comes it by harbouring in its rumen a culture of microorganisms which assist the process of digestion. The food is retained for a considerable period while a complex set of fermentation processes takes place, resulting in the conversion of the structural tissues to soluble carbohydrates. In principle, this process is not very different from that occurring in a detritus food web. One is in the environment, the other is in a culture within the animal. Consider also that in many parts of the world with low rainfall such as Australia, or the prairies, ruminants feed for long periods on brown, dead plant material and it seems as if the distinction between detritus food webs and grazing food webs is tenuous.

Some recent work in Nova Scotia has shown that sea urchins can feed equally well on living or dead macrophyte material, or on diatoms when deprived of macrophytes. Moreover, they have a capacious gut which has a bacterial culture of about the same population density as in the rumen of a cow, and the bacteria are capable of digesting cellulose and carrying out nitrogen fixation. Rather loosely, one may call these urchins 'the ruminants of the sea'.

Furthermore, a study of a mysid shrimp which lives in dense beds of *Zostera* has shown that it has a capacity to digest cellulose, but when treated with antibiotic solution it loses that capacity. To show that the shrimp itself was unharmed, it was reinoculated with microorganisms and the ability to digest cellulose was restored.

There is no doubt that zooplankton graze phytoplankton directly, and rely on them as their major source of food, especially in the open ocean. When phytoplankton numbers increase, zooplankton numbers also increase after a short delay, and vice versa. Hence we have evidence of a true grazing food chain, utilizing fresh plant tissue directly. However, a proportion of the phytoplankton sinks and decays on the bottom, entering detritus food webs (Chapter 5), while another fraction passes undigested through the guts of zooplankton and sinks as faecal pellets (Chapter 3). Hence, there are strong connections between grazing and detritus food webs in the plankton, and it is possible that even in aquatic systems, more energy is channelled through detritus-based systems than through grazers (Chapter 5).

10.3   Long and short food chains

For convenience in this section we shall discuss linear food chains as if they existed in nature. In fact, as we know, most natural systems have food webs, and a linear food chain is only a convenient abstraction. The transfer of energy along a food chain involves losses of the order of 80 to 90% at each stage. Thus, after transfer through four levels, as in phytoplankton–zooplankton–herring–cod, 1000 calories of phytoplankton will yield only 1–8 calories to the cod. Hence it is an ecological advantage to keep the number of steps in a food chain to a minimum. Why, then, does not a cod feed directly on phytoplankton? The answer is that it would expend more energy in catching such small, dispersed particles than it gained from the food. It appears that most animals are equipped to take food organisms that are about 1% of their own body weight. There are exceptions: whales have a very special apparatus for straining small food organisms out of huge volumes of water, but in general the 1% rule appears to hold. For an animal which is taking food at the lower size limit for making a living, it helps if the food is very abundant. Hence, there are two factors which determine the length of a food chain: how large are the particles of food at the primary producer level, and how abundant are they? Fig. 10.2 shows how the efficiency of transfer from primary to secondary production in various upwelling systems depends on the intensity of primary production.

There is evidence that phytoplankton species growing in more productive marine environments tend to be larger than those growing in oligotrophic situations like the open ocean. For the latter, small size is an advantage in

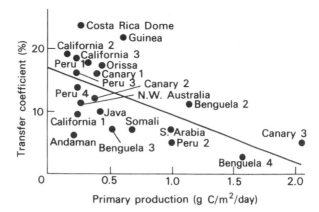

Fig. 10.2. Relationships between transfer coefficients and productivity. [From Cushing 1971b]

providing a greater surface-to-volume ratio which delays sinking and improves nutrient uptake. Hence, food chains in the open sea start with small cells, too small for ingestion by most copepods. The next step in the food chain may be a protozoan, the next a small copepod, and so on. Moreover, since organisms are not abundant, the size difference between predator and prey tends to be smaller. By contrast, in estuaries and upwelling areas, primary production is often dominated by fairly large diatoms which are ingested by zooplankton or even small fish, so the food chains are shorter. Ryther (1969) suggested that the average length of a food chain in the open ocean was five steps from primary producer to harvestable fish, whereas in coastal waters it was three steps and in upwelling areas one or two. He suggested that transfer efficiencies were higher in the more productive areas, and used the estimates to calculate the potential yield of fish from the open ocean, coastal zone and upwelling areas (Chapter 7). The conclusion (Table 7.2) was that coastal and upwelling areas each account for nearly 50% of the world's fish production, and that the open ocean contributes a trivial amount. The details of the calculation are open to criticism, but the conclusion is probably valid.

A point to be borne in mind when discussing food chains is that all organisms pass through a succession of size classes during development. For example, young cyprinid fish in European rivers can feed directly on protozoa in the detritus food web. Hence, the size relationships in the food webs are constantly changing with time, and we must be on our guard against oversimplification.

In any calculation involving discussions of the density of a food species, we must be careful not to consider only the average density. It is well known that planktonic organisms tend to occur in patches (p. 41), and it has been shown several times that various zooplankton species can only find adequate amounts of food by concentrating their feeding activities on the dense patches.

10.4 Factors influencing diversity and stability

In both plankton and benthos we have seen that species diversity tends to increase during periods of environmental stability. In times or places characterized by unstable or changing environmental conditions, the communities are made up of a relatively small number of *r*-strategy species, but when conditions are stable numbers of organisms increase, resources become limiting, and these conditions favour K-strategy species able to maintain maximum biomass while using the minimum of resources. According to the stability-time hypothesis of Sanders (Chapter 6), long periods of stable conditions enable more and more species to co-exist, each one becoming biologically accommodated to species occupying similar niches to itself.

We should note that the time scale of events to which Sanders was referring is of the order of hundreds of years, yet similar increases of species diversity are observed during periods of stability in the plankton which may last only a few weeks. Since the generation time of planktonic organisms is shorter, we should expect some compression of events, but this difference does seem to be rather extreme. Is it possible that the stability-time hypothesis is in need of further development? Some ideas on this subject have come from intensive study of marine intertidal communities. The life histories of the invertebrates found there are longer than those of most planktonic animals but shorter than those in the deep benthos. Moreover, the intertidal animals are conveniently accessible for experimental manipulation.

A good discussion of recent ideas was given by Menge and Sutherland (1976). They drew attention to two important factors governing species diversity: competition, and predation. They pointed out that if a large number of species at the same trophic level are left to co-exist for a long period in a stable environment, each population will build up to the point where resources are limiting and will be in severe competition with other species. At this stage, there are two conflicting theories about what will happen. The 'compression hypothesis' states that interspecific competition selects for increased specialization, in which each species reduces the array of habitats or resources which it uses. This may pave the way for successful invasion by additional species.

An alternative view is that under intense competition some species will be more successful and tend to eliminate others by the process of competitive exclusion. This would lead to decreased species diversity, which empirically is not what we observe. This is where predation comes in to modify the competitive process. Menge and Sutherland maintain that competitive exclusion occurs only when species are allowed to reach maximum population density (K). If predators are active and succeed in keeping prey at levels below the limit of resources, competitive exclusion will be alleviated, and greater species diversity will result.

Menge and Sutherland (1976) have therefore suggested that an alternative interpretation of Sanders data would be that his deep-sea faunas of high-species diversity contain more species of predators and more abundant predators. If this is true, they would have the effect of keeping prey density below carrying capacity, and reducing competitive exclusion. Menge and Sutherland have also drawn attention to differences between trophic levels. It is not surprising to find that herbivores have predators, but the situation in which those predators are acted on by a still higher level of predators is not common. Hence, at the predator level competition is more likely to be an important regulating factor. By extrapolation, the theory is put forward that in communities with larger numbers of trophic levels, predation becomes more important as an organizing factor, and leads to greater species diversity. It has been observed many times that stable or highly predictable environmental conditions lead to the development of longer food chains, and that conversely increase in environmental instability leads to the elimination of more specialized consumers of higher trophic status. Menge and Sutherland suggest that the explanation of low-species diversity in highly fluctuating or unpredictable environments is the scarcity of predators and the operation of competitive exclusion at lower trophic levels.

Margalef (1968) and Odum (1969) expressed a similar idea when they observed that 'mature' ecosystems have large numbers of 'K' species with low $P:B$ ratios, while early stages in the maturation process favour 'r' strategists with high $P:B$ ratios. If we interpet 'mature' as meaning that the ecosystem has persisted through a considerable period of constant or highly predictable environmental conditions, and that it has developed a structure which includes several trophic levels (long food chains), then under these conditions the theory of Menge and Sutherland predicts that predation will structure the community and lead to high species diversity.

The common ground running through the debate is the idea that in the relatively constant conditions of the deep lakes and oceans or in tropical waters with a permanent thermocline, there is high species diversity, and that environmental constancy is the cause of system diversity and stability. The older idea that diversity itself generates stability now has very little support (see p. 109).

In this respect, terrestrial systems seem to parallel aquatic systems. In a discussion of 'why the earth is green', Hairston et al. (1960) suggested that land plants are seldom grazed to extinction, yet there are vast numbers of potential herbivore populations that could do so. They concluded that herbivores as a whole are not food limited, but are held well below the carrying capacity of the food resource by their predators. Hence, predator control of lower trophic levels may be the characteristic feature of both terrestrial systems and of the more constant aquatic systems.

10.5   The overall pattern of energy and nutrient flow

10.5.1   RIVERS

Early work on patterns of energy flow in aquatic systems was carried out in springs, where temperature and hence the metabolic rate of organisms tends to be relatively constant year-round. From these, it was a relatively easy step to begin work in rivers, where the water is well mixed and shallow, compared with most lakes. Fig. 10.3 shows the pattern of energy flow in Silver Springs, Florida, a very large spring where the outflow is a river approaching 100 m wide in places. The macrophyte *Sagittaria* produced nearly 3000 kcal m^{-2} yr^{-1} but the periphyton on its surface produced even more, around 6000 kcal m^{-2} yr^{-1}. Production of herbivores amounted to nearly 1500 kcal m^{-2} yr^{-1} indicating an efficiency of energy transformation of about 16%, but carnivores had an overall efficiency around 5%. It is a characteristic of river ecosystems that they receive considerable inputs of allochthonous organic material from the basins in which they lie, and this supplements the organic matter synthesized in primary production in the river (autochthonous). Successive examples in Fig. 10.3 show proportionately greater inputs of allochthonous matter. The extreme case is Bear Brook, a heavily shaded woodland stream with almost no autochthonous

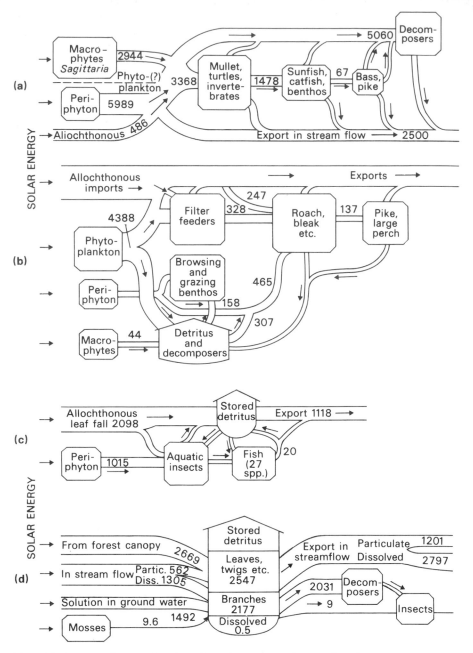

Fig. 10.3. Patterns of energy flow in various river systems (kcal m^{-2} yr^{-1}). (a) Silver Springs, (b) River Thames; (c) New Hope Creek (sta. 6); (d) Bear Brook. [From Whitton 1975]

production, receiving almost all its input from a forest canopy. In this stream, bacteria and fungi reduced the detritus to dissolved organic compounds, but there was negligible production of invertebrates or fish.

Of the examples shown, only the River Thames had significant amounts of phytoplankton production. This river has a very long history of human navigation, and now has a dam (or weir) and a ship lock every 5–10 km, to maintain constant depth for navigation. Biologically it is therefore a series of impoundments. It also receives large volumes of suspended solids and dissolved nutrients from sewage effluents, and is heavily eutrophicated. In general, only large slowly-moving rivers have significant phytoplankton production. Even then, it is not entirely clear how populations maintain themselves against a current. It is assumed that there is enough still water in eddies and backwaters to 'seed' the flowing water with phytoplankton cells. The most characteristic primary production in rivers is by macrophytes and periphyton. Production figures of 1000 kcal m^{-2} yr^{-1} ($= 100$ gC m^{-2} yr^{-1}) are not uncommon. The production of 2944 kcal m^{-2} yr^{-1} by *Sagittaria* in Silver Springs is one of the highest on record. It must be remembered that in most running water situations periphyton and macrophytes are confined to the fringes, and the figures given above are averaged over the whole width of the river.

Since rivers are in such intimate contact with their river basin, receiving supplies of nutrients both in tributaries and in seepage water, it is extremely difficult to follow patterns of nutrient turnover. In an experiment in which ^{32}P was dripped in to a stream, it was found that the phosphorus was taken up by periphyton in a matter of minutes. Consumers were labelled in a few days, and the label moved slowly downstream as it was cycled several times through the trophic levels.

10.5.2   LAKES

A constantly recurring theme in this book is the analysis of aquatic ecosystems in terms of the flow of energy and the cycling of nutrients. Measurements of primary and secondary productivity, when expressed in energy units, are excellent indications of the flow of energy between trophic levels. In early, pioneering studies of lakes no clear distinction was made between biomass and production; a lake with a large standing stock of plants and animals was described as productive, even though there were no production data to support the statement. In the 1960s a cooperative effort, the International Biological Programme (IBP) had as one of its aims the collection of productivity data, which would help clarify the relationship between P and B. At the same time, in many lakes, records were kept of the changing concentrations of nutrients, especially nitrogen, phosphorus and silicon. When the time came to analysing the results from 51 lakes scattered across the globe, Brylinsky and Mann (1973) asked the question "Which factor gives the best correlation with primary productivity: solar energy flux or nutrient concentration?" They used latitude as an index of solar radiation, and found that the correlation with latitude was much better than the correlation with any nutrient concentration (Fig. 10.4).

To the majority of limnologists this seemed a very strange result. It is well known that addition of nutrients, especially phosphorus, to a lake leads to increased primary productivity, often to the point of making the lake an unattractive green 'soup'. As Brylinsky and Mann (1973) pointed out, it is the rate of *cycling* of nutrients that is important, not the *concentration* in the water at any one time. Subsequent work has confirmed this. When the flux of nutrients into lakes from the catchment area is measured, there is a good correlation

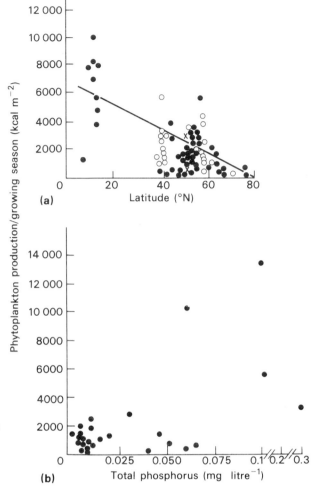

Fig. 10.4. Relationships of gross primary productivity in 51 lakes studied under the IBP, to (a) latitude (as an index of solar energy input) and (b) phosphate in the water.

between nutrient flux and primary productivity. The distinction between the energy (in production) and the amount present (in biomass) is now well recognized, but the distinction between nutrient flux, or turnover, and nutrient concentration is still being overlooked. It normally requires the combination of two sets of data: flow of water and concentration of nutrients in that water. A competent study of the functioning of aquatic ecosystems requires skill in hydrology as well as biology and chemistry.

The IBP results enabled Brylinsky and Mann (1973) to look at ratios between productivities at various trophic levels (Table 10.1). For technical reasons it was necessary to work with gross primary productivity, i.e. estimates of the total energy fixed by the phytoplankton, rather than the net amount available to consumers. The figures ranged from less than 500 kcal ($= 50$ g C) m^{-2} yr^{-1} in arctic lakes to over 10 000 kcal ($= 1000$ g C) m^{-2} yr^{-1} in some tropical lakes. The average efficiency of conversion of incident solar energy was 0.4%. The ratio of herbivore production to gross photosynthesis was 13.7% in the plankton

Table 10.1. Efficiency with which solar energy was transformed to plant gross production, and efficiency with which this was transformed to secondary production. [From Brylinsky and Mann, 1973]

Trophic level	Number of lakes	Mean (%)	Range (%)
Phytoplankton	55	0.4	0.02–1.7
Herbivorous zooplankton	21	13.7	2.6–21.2
Carnivorous zooplankton	12	1.4	0.2–2.5
Herbivorous benthos	11	7.5	0.5–9.5
Carnivorous benthos	10	0.9	0.1–3.4

and 7.5% in the benthos. The ratio of carnivore production to gross photosynthesis was 1.4% in the plankton and 0.9% in the benthos. Since benthic animals tend to live longer they divert more energy to maintenance of biomass, and it is therefore to be expected that they will be less efficient than the plankton. If the calculations had been made on the basis of net primary productivity, the efficiency of the planktonic herbivores would have been about 17%. Results from marine environments also suggest that zooplankton have efficiencies approaching 20% (see below).

10.5.3 THE SEA

To illustrate energy flow characteristics, two examples will be taken: (a) the North Sea to show the broad relationships between all trophic levels in a system that for the purposes of discussion is regarded as self-contained, and (b) a fringing subtidal community in Canada to show how the details of energy flow can be charted at the species level, but also illustrating a situation in which there is a massive import of nutrients and export of plant production.

An energy flow diagram for the North Sea was put together very tentatively by Steele (1974). In Fig. 10.5, his data are rearranged to facilitate comparison with Fig. 10.3. All fluxes are in kcal m^{-2} yr^{-1}, and the evidence for the values used is as follows. Phytoplankton production in the North Sea averages 900 kcal (90 g C) and this is grazed very efficiently by the zooplankton, giving rise to 300 kcal of faecal pellets which sink to the benthos. Pelagic fish prey on zooplankton, and their biomass and food requirements are fairly well known, suggesting that they take about 85 kcal from the zooplankton. Invertebrate predators such as *Sagitta* and ctenophores probably take an equal amount. The quantitative role of the bacteria in relation to meio- and macrobenthos is still an open question (Chapter 6) but from known biomass data and estimated turnover rates it is thought that the benthos produces about 50 kcal. Of this, 30 kcal is required by the demersal fish stocks, but another 20 are probably taken by the benthic carnivores, such as shrimps, crabs and lobsters. Yields to man are about 4 kcal of pelagic fish and 2 kcal of demersal fish, but it must also be recognized that demersal fish take about 4 kcal of pelagic fish and 2 kcal of benthic predators.

In order to balance the budget, or bridge the gap between primary production and fish yields, it is necessary to postulate that the efficiency of zooplankton

195

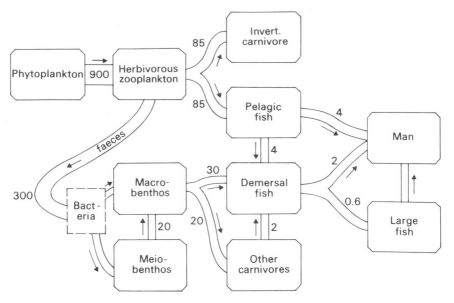

Fig. 10.5. Pattern of energy flow in the North Sea (kcal m^{-2} yr^{-1}) from data put forward tentatively by Steele (1974). Nutrients are regenerated from every compartment, but zooplankton and benthos are the most important.

is close to 20%. The amount of energy input to the macrobenthos is not known, owing to our ignorance of the efficiencies of bacteria and meiobenthos in transforming energy. We can only observe that the benthos as a whole needs to have an efficiency of 16 or 17%, and the macrobenthos an even higher efficiency.

The North Sea has been fished more intensively and for a longer period than almost any area in the world. During the last few decades there have been marked changes in species abundance so that, for example, herring catches have declined while mackerel catches have increased in a different area. The overall view given by the energy flow diagram helps us to put these changes in perspective. So long as there is phytoplankton and zooplankton production it is likely that pelagic fish populations will be there to utilize the production. Changing species patterns cause hardship in industries geared to catching particular species (such as herring) but the system continues to function with the same overall pattern.

Our second example (Fig. 10.6) is taken from the subtidal fringing community on a rocky bottom in a Canadian marine bay with an area of some 140 km². The seaweeds, dominated by *Laminaria* had an extremely high productivity, almost ten times that of the phytoplankton. Sea urchins living in the area consumed only about 3% of this production, while other invertebrates took even less. The balance was exported from the fringing community to communities in deeper water. In return the fringing community received large amounts of nutrients regenerated in deeper water. Sea urchins were the chief secondary producers in the area, and were preyed upon by lobsters. It was shown experimentally that lobsters prefer crabs as prey, with urchins being their second preference. However crabs are much less abundant and are difficult to catch, so the lobsters ate large numbers of sea urchins.

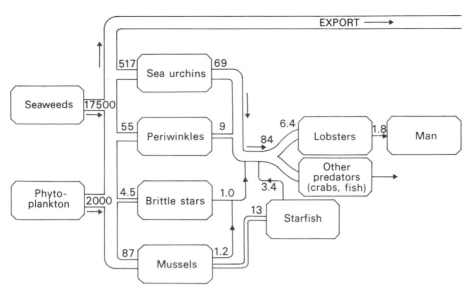

Fig. 10.6. Pattern of energy flow in a fringing system. Note that there is a massive export of organic detritus to adjoining systems. There will be a massive return of nutrients to permit the continuation of primary production. [From data in Miller *et al.* 1971]

There is a sequel to this story. Under intense fishing pressure lobster stocks declined. Then the sea urchins formed dense local aggregations and began to overeat their food supply, until the kelp beds had been destroyed over most of the area they had previously inhabited. This led to a community with a much lower productivity and a different community structure. The new configuration has persisted for 10 years and spread over hundreds of miles of coastline. The strange, and alarming feature of this situation is that the sea urchins which had been the passive consumers of only 3% of the primary production could, by aggregating and attacking the *Laminaria* plants one at a time, effectively destroy the source of primary production in a few years. The evidence is incomplete but it looks as if the event was triggered off by removal of the key predators, the lobsters. It is an extreme example of a predator being responsible for structuring a community.

The high productivity in Fig. 10.6 is characteristic of many fringing communities. Salt and freshwater marshes, seagrass beds, kelp beds, and coral reefs are all examples of fringing communities in which carbon fixation is of the order of $1000 \text{ g C m}^{-2} \text{ yr}^{-1}$, and energy fixation about $10\,000 \text{ kcal m}^{-2} \text{ yr}^{-1}$. A common feature of these systems is the arrangement in which macrophytes are anchored in one place while water moves over or between them, bringing a continuous, if dilute, supply of nutrients. Another characteristic is a tendency to recycle nutrients very efficiently within the system. In coral reefs, for example, unicellular algae live within the tissues of the secondary producers, and take up nutrients before they have been excreted. Many salt marsh plants have nitrogen-fixing bacteria living in and around their roots, and it has been found that they fix gaseous nitrogen from the water when the ammonia concentration is low, but reduce nitrogen-fixing activity when ammonia is more abundant.

197

It was stated near the beginning of this chapter that the flux of nutrients operates the gate, or valve, that controls the flow of energy. In systems language this means there is a feedback in the system. The flow of energy through the plankton and benthos is associated with nitrogen and phosphorus excretion, and these nutrients, when fed back to the primary producers, make possible further energy flow in primary production. An example of this interaction can be given from the *Laminaria* fringing community described in Fig. 10.6. The growth rate of the seaweeds accelerated through the winter when nitrate was abundant, reaching a peak just at the time when the sea became stratified and the phytoplankton began to reduce the nutrient concentrations in the surface water. In the summer the growth of the *Laminaria* declined, in spite of rising temperature and an increase flux of solar energy. Fertilization of the water with nitrate immediately led to a burst of seaweed production, showing that lack of nitrate was limiting primary production. Similar results have been obtained with phytoplankton, indicating that in general the growth of algae in the sea is limited by the availability of nitrogen.

In many lakes, however, phosphorus appears to be the key element limiting primary productivity. It has been observed many times that addition of phosphorus-rich sewage pollution to a lake ecosystem leads to increasing primary production, and that the effect is due to changes in phosphorus rather than other nutrients. Conversely, when sewage effluent was diverted from Lake Washington, near Seattle, U.S.A., there was a rapid and parallel drop in phosphorus and phytoplankton biomass, while nitrate dropped less rapidly (Fig. 10.7). When a smaller lake only 20 km from Lake Washington was given similar treatment it failed to show the desired response within 5 years. Lake Washington is deep and does not develop an anaerobic hypolimnion, but the second lake develops an anaerobic hypolimnion, with reduction of ferric ion and release of adsorbed phosphate. In a lake of this type the sediments are more important than inflow in determining the availability of phosphate (see Chapter 6).

10.6   A systems-analysis view of aquatic ecosystems

Traditional biology, in its growth from natural history, has tended to place the organism in centre-stage as the unit of biology. Evolutionary biology treats the individual as the unit of natural selection, and taxonomy defines the species by reference to individual type-specimens. When physics and chemistry were brought to bear on the problem of 'how an organism works', physiology tended at first to be the physiology of organisms, later it became possible to work on organs, tissues and cells in isolation.

There are still many biologists who approach an aquatic ecosystem with questions such as 'Which organisms live here, and how do they relate to their environment?' However, this book has shown that in addition to questions about organisms and populations of organisms, there is a whole suite of questions about interactions and processes which are not so much the properties of organisms as properties of the ecosystem. Fig. 10.7 gives data on nitrate,

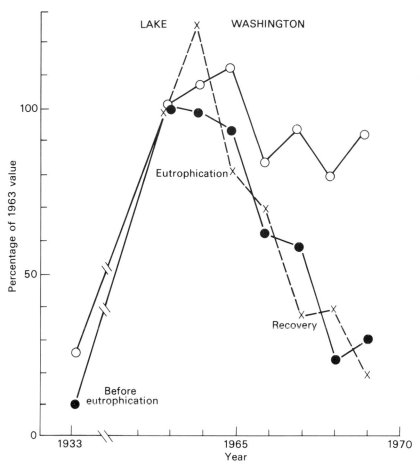

Fig. 10.7. Phytoplankton biomass (expressed as chlorophyll), nitrate and phosphate in Lake Washington, taking 1963 values as 100%. ○ Nitrate. ● Phosphate; × Phytoplankton biomass (chlorophyll). [From Edmondson 1970]

phosphate and chlorophyll, none of which are the properties of an organism. They pertain to the system.

Systems science deals with just such phenomena. It recognizes that systems have properties which are distinctive and are not to be understood by summing the properties of the components. Systems have *levels of organization*, each of which has its own characteristic phenomena. Cells, tissues, organs and organisms are levels of organization which are familiar to us. It would not be sensible to try to predict the behaviour of, say, a fish by making studies only at the cellular level. It is necessary to observe the fish behaving at its own organismal level of organization. Similarly, it is necessary to discuss, observe and measure the properties of whole aquatic ecosystems if we are to understand them and make predictions about them. Energy flow diagrams are ways of representing the 'anatomy' of these systems, and this can be done at various levels of resolution from the organismal (Fig. 10.6) to the broad functional group (Fig. 10.5). However, the functioning of these systems cannot be properly understood from energy flow alone. We have to combine it with knowledge of nutrient cycling.

And in aquatic systems we also need hydrological information to enable us to evaluate upwelling processes and import–export patterns of the system of interest. This view of the world is a young one, and very incomplete. There is much to be done, and many questions which as yet we do not even know how to ask.

Up to now, the most useful tool that ecologists have been able to borrow from systems science is the simulation model. For example, starting with an energy flow diagram, it is possible to write equations that govern the transfer of energy from one compartment to another, and after setting the starting values for each compartment, allow a computer to calculate and print out a time series for the values in each compartment. These can be read as the changes with time that would be expected in an ecosystem as a consequence of the assumptions made when writing the equations. It is a powerful tool for examining complex interactions and has been used for example, to make predictions about the consequences of adding sewage pollution to an estuary. Those wishing to find out more about simulation modelling should refer to Patten (1971, 1972, 1975).

11

Aquatic Systems as Part of the Biosphere

G.M. WOODWELL

11.1   Introduction

The biosphere is that portion of the earth that supports life. It extends a few millimetres into the sediments of the abyssal depths of the oceans and to the tops of the earth's highest mountains. Spores carried by winds extend it further to heights of 30 000 feet and beyond into the stratosphere but the normal range of living systems is restricted to lower elevations. The modern biosphere is not only the product of evolutionary processes operating over at least three of the earth's 4.5 billion year history, but is today maintained in a surprising degree by the intricate processes of life in interaction with complex physical and chemical systems. The continuing and exciting explorations of the other planets of the solar system, including the moons of Jupiter, show that the earth is unique: there is no other planet with a biosphere in our solar system, no other planet with lakes, rivers and oceans, no other planet that could support life.

Although the emphasis of this book is on aquatic systems, there is such an intricate series of interactions between aquatic and terrestrial systems at the level of the biosphere that we must consider both if we are to gain insights into either. Many factors are involved in maintaining the biosphere as a place suitable for life and especially for man, and it is easy to show that if any one of them were modified, the biosphere would be grossly different. The factors extend from life itself to the influx of solar energy as light and heat; they include the albedo or reflectivity of the earth, the fact that about two-thirds of the surface of the earth is water, and that the continents occupy certain positions relative to the axis of rotation and the equator. All of these factors affect climate worldwide and therefore bear directly on the capacity of the earth for supporting people, a topic that is coming increasingly to the fore as human population climbs to press all resources toward their limits or beyond and small changes in environment gain greater importance in the human lot. A glance at a map of the earth will show that a small warming of the earth due to a change in the energy output of the sun, a change in the earth's albedo, or other factors would shift the arid zones poleward to cover much more of the land area than they do now. Such a change would have profound implications for food production and the habitability of the earth by man.

The details surrounding the evolution of the biosphere remain obscure, but the evidence is overwhelming that the relationships between the evolution of life and the environment are reciprocal. It seems clear also that the general conditions for life have remained the same for a long time, perhaps a billion

Table 11.1. Primary production and biomass estimates for the biosphere.*

1	2	3	4	5	6	7	8
Ecosystem type	Area, 10^6 km^2 = 10^{12} m^2	Mean net primary productivity, g C/m^2/year	Total net primary production, 10^9 metric tons C/year	Combustion value, kcal/g C	Net energy fixed, 10^{15} kcal/year	Mean plant biomas, kg C/m^2	Total plant mass, 10^9 metric tons C
Tropical rain forest	17.0	900	15.3	9.1	139	20	340
Tropical seasonal forest	7.5	675	5.1	9.2	47	16	120
Temperate evergreen forest	5.0	585	2.9	10.6	31	16	80
Temperate deciduous forest	7.0	540	3.8	10.2	39	13.5	95
Boreal forest	12.0	360	4.3	10.6	46	9.0	108
Woodland and shrubland	8.0	270	2.2	10.4	23	2.7	22
Savanna	15.0	315	4.7	8.8	42	1.8	27
Temperate grassland	9.0	225	2.0	8.8	18	0.7	6.3
Tundra and alpine meadow	8.0	65	0.5	10.0	5	0.3	2.4
Desert scrub	18.0	32	0.6	10.0	6	0.3	5.4
Rock, ice, and sand	24.0	1.5	0.04	10.0	0.3	0.01	0.2
Cultivated land	14.0	290	4.1	9.0	37	0.5	7.0
Swamp and marsh	2.0	1125	2.2	9.2	20	6.8	13.6
Lake and stream	2.5	225	0.6	10.0	6	0.01	0.02
Total continental	149	324	48.3	9.5	459	5.55	827
Open ocean	332.9	57	18.9	10.8	204	0.0014	0.46
Upwelling zones	0.4	225	0.1	10.8	1	0.01	0.004
Continental shelf	26.6	162	4.3	10.0	43	0.005	0.13
Algal bed and reef	0.6	900	0.5	10.0	5	0.9	0.54
Estuaries	1.4	810	1.1	9.7	11	0.45	0.63
Total marine	361	69	24.9	10.6	264	0.0049	1.76
Full total	510	144	73.2	9.9	723	1.63	829

*All values in columns 3 to 8 expressed as carbon on the assumption that carbon content approximates dry matter $\times 0.45$.

isotope of carbon, ^{14}C, under experimental conditions. The technique is open to substantial error (see Chapter 2) and probably does not provide data that are strictly comparable to the data for terrestrial systems. There is, nonetheless, a sufficient body of these data from the combination of these techniques to allow a worldwide appraisal of the primary production of major segments of the biosphere. The most widely-used such appraisal is that of R.H. Whittaker and G.E. Likens, published several years ago. It is reproduced here in modified form as Table 11.1.

The total amount of carbon fixed worldwide according to this estimate is about 75 billion metric tons. Of this about 25 billion metric tons is fixed in aquatic systems. Most of the remainder is fixed in forests.

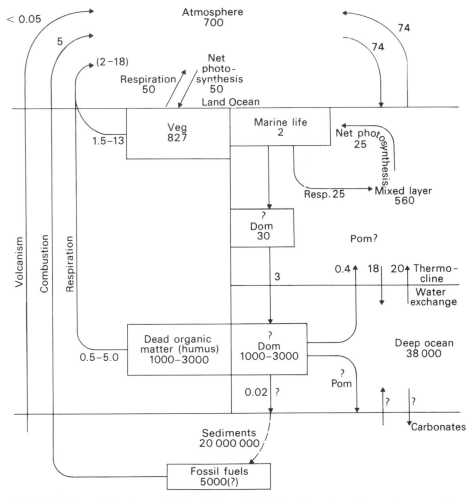

Fig. 11.1. The world carbon cycle. The CO_2-content of the atmosphere is increasing annually by about 2.3×10^{15} g due to the excess of releases of CO_2 into the atmosphere over removals. The major reservoir of carbon in exchange with the atmosphere is the ocean, but the rate of transfer of carbon into the deep ocean is slower than the rate of release into the atmosphere from combustion of fossil fuels and from destruction of the biota. Units are 10^{15} g (10^9 metric tons).

205

The relative importance of the various segments of the biota in the circulation of carbon through the biosphere can be seen by reference to Figure 11.1. The diagram contains the best current estimate of the amount of carbon held in various parts of the biosphere. The most important comparison for us at the moment is the comparison between the total standing crop of carbon held in the biota, about 830×10^9 tons (Table 11.1), and the total amount estimated to be in the atmosphere, about 700×10^9 tons. The amount of carbon fixed annually by green plants worldwide is thought to be in excess of 10% of that held in the atmosphere. This relationship simply emphasizes the potential importance of the biota in affecting the carbon dioxide content of air worldwide in the short-term. There is, moreover, special importance to be given to forests because of the magnitude of the pool of carbon that they contain and the fact that they carry out at least 50% of the net production of the biosphere.

The importance of these relationships is seen best through an analysis of data on CO_2 concentrations in air collected by C.D. Keeling and his colleagues of the Scripps Institution of Oceanography of the University of California. The data were taken on Mauna Loa, a volcanic peak in the Hawaiian islands (Fig. 11.2) and show two important changes in the atmospheric CO_2 since 1958

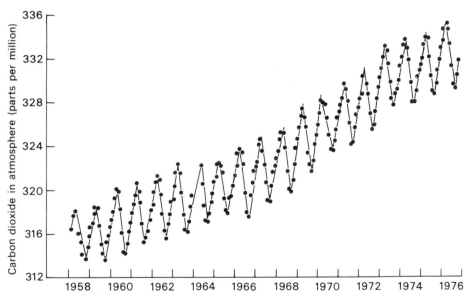

Fig. 11.2. The record of CO_2 in the atmosphere at the Mauna Loa Observatory in the Hawaiian Islands. Data were taken by C.D. Keeling and his colleagues beginning in 1958.

when the record was started. First, there is a year-by-year increase in the CO_2-content of air that averages over the 20-year period about 1 ppm annually. The CO_2 content in 1979 was about 336 ppm by volume. Second, there is an annual oscillation in the CO_2-content that has an amplitude of about 5 ppm at Mauna Loa. The oscillation is thought to be caused by the metabolism of temperate zone forests.

The Mauna Loa data are but one of many observations that show a similar worldwide pattern of increasing CO_2 in air coupled with a seasonal oscillation in the CO_2-content. The peak in the seasonal oscillation occurs in late winter and the minimum in late summer. The year-by-year increase in CO_2 is due in part to the combustion of fossil fuels, now thought to release annually about 5 billion metric tons of carbon as CO_2 into air. It is also due to the release of CO_2 from the destruction of forests and the oxidation of humus. The magnitude of this release is not known with precision; it is estimated as 4–8 billion metric tons, approximately equivalent to the fossil fuel release.

Part of the CO_2 is obviously accumulating in the atmosphere, but the total quantity accumulating, about 2.3×10^9 tons, is substantially less than the total released. The remainder is sequestered elsewhere, probably in the oceans.

The details of this important cycle can be seen best by examining the diagram of Fig. 11.1. It is important in considering this figure to separate the size of the pools of carbon that are interacting from the rates of exchanges between the pools. First, it should be apparent that the oceanic pool is very large in proportion to the atmosphere's 700×10^9 tons and the biota's 830×10^9 tons.

The carbon in the oceans is present in several places and forms, and is not all readily available for exchange with the atmosphere. By far the largest amount of carbon in the oceans is held in solution as dissolved carbon dioxide, which enters into a series of complex reactions to become a part of the carbonate-bicarbonate system. The total amount of carbon held in this system is estimated as about $38\,000 \times 10^9$ metric tons in the deep waters of the oceans below the thermocline and about 560×10^9 metric tons above the thermocline. There is in addition a large total amount of carbon held as dissolved organic matter (d.o.m.) in the oceans, but its concentration is very low, about 1 ppm or 1 mg/litre. The total is large, $1000–3000 \times 10^9$ metric tons, because the d.o.m. occurs throughout the enormous volume of the oceans. Most evidence indicates that the d.o.m. is not very active biotically. The carbonate sediments of the ocean bottom contain a very large amount of additional carbon, most of which is not immediately available for mobilization unless there is a change in the acidity of the oceans. By comparison with these levels of carbon, the amount held in the biota of aquatic systems is small indeed, probably less than 3×10^9 metric tons worldwide.

The capacity of the oceans for absorbing carbon from the atmosphere is nominally very large indeed; the problem is the time required for the exchange. The time for exchange between the abyssal waters and the mixed surface waters through mechanical mixing is commonly considered to be a thousand years or more, and it is here in the deeper waters of the ocean, not in the mixed layer, that the large capacity for storing atmospheric CO_2 resides.

The second part of the problem is to determine the rate of transfer of CO_2 between the atmosphere and the various other segments of the biosphere that are in interaction with the atmosphere. The arrows in the diagram show the best estimate of these rates.

The magnitude of the total exchange between the atmosphere and the surface oceans is large, about 74×10^9 metric tons. This total is the total amount of carbon that diffuses as CO_2 into the surface of the ocean annually from the atmosphere. There is of course a movement of carbon in the opposite direction, also by diffusion. The rate of diffusion into the oceans apparently exceeds the rate of movement back into the atmosphere by a small amount, currently thought to be 2–3×10^9 metric tons annually. The difference is the amount transferred from the mixed layer of the oceans into the depths by mixing and by sedimentation of organic matter. The mixing process is much more important because so much more carbon is held in the inorganic system than in the organic form. The net exchange is obviously too small to prevent the current rise in the CO_2-content of the atmosphere.

One might wonder if the approximately 25×10^9 metric tons of carbon available in the oceans annually in net primary production might result in the rapid transfer of carbon to the abyss where it becomes either a part of the inorganic carbon pool or is sedimented. The answer appears to be that any such storage would have to have been stimulated in some way in direct proportion to the rate of mobilization of fossil fuels. Such a transfer would require an increase in the amount of nitrogen and phosphorus available, according to the Redfield ratios, and such an increase is not yet measurable and seems not to have occurred.

At the moment there is no satisfactory basis for predicting the future CO_2-content of the atmosphere beyond the mere projection of the current trend shown in the Mauna Loa and other data. The knowledge of the mechanisms by which CO_2 is removed from the atmosphere, and added to it, is too crude to allow a more refined model to be useful.

The topic is important because the increase in CO_2 in air is thought to have the potential for causing a general warming of climate in the next 2–3 decades. Such a disruption of the biosphere is obviously undesirable when all resources are being used at their limits and any disruption is almost certain to cause at least a temporary reduction in essential supplies for support of man.

11.3   The nitrogen cycle in aquatic systems

Nitrogen, phosphorus and potassium are the three 'fertilizer elements', normally added to agricultural soils to assure maximum yields of crops. The basic theory that applies is Liebig's Law of the Minimum: growth is limited by the factor that is available in the minimum quantity. The law is useful in agriculture and is commonly thought to apply to nitrogen in particular because an increase in growth can usually be obtained throughout a very wide range of applications of nitrogen. There is a tendency to use the same logic in analysis of natural communities, especially in aquatic systems, but this use of the theory is misleading.

One of the central principles of ecology is that evolution tends to eliminate individual limiting factors. The process is of course not planned, deliberate, or part of any strategy; it is merely the inevitable result of the same series of random events that over a very long time produce species, communities, and ecosystems. Evolution tends to develop systems that exploit whatever resources are available. The process results over time in the division of resources among different species and among groups of species. As the diversity of populations increases, the opportunities for mutualism increase and the community itself is obviously adjusted progressively to the array of physical, chemical and biotic resources available at any time. Thus while Redfield observed that the average ratio of $P:N:C$ in marine communities is $1:16:106$, he recognized that individual species deviate from this mean substantially.

If we were to go one step further and alter the ratio of nutrients available in the water, we would see changes in the community as different populations of species are formed. The effect, however, would be in the direction of the well-established patterns of eutrophication; not the stimulation of growth of the normal array of species in the community, but the displacement of those by others better suited to the altered chemistry of environment. The response to enrichment by nutrients is not the enhanced growth seen in agriculture where fertilizer is applied but a complete change in the community itself. Obviously the concept of limiting factors does not apply in these communities in the same way it applies in agriculture: the shift in nutrients changes the array of plants. In aquatic systems we define these changes in their early stages as 'eutrophication'. In later stages we recognize the intensified pattern as 'pollution', although pollution may have other causes. Despite the importance of agriculture in the world economy the biotic segment of the world nitrogen cycle is dominated by natural systems whose species composition is sensitive to changes in the ratios of nutrients available.

The availability of nitrogen in the biosphere is obviously one of the most important factors affecting the biota. While its effect is probably not to be considered that of a limiting factor, changes in availability of nitrogen share with a few other factors, including phosphorus, toxins, light, and temperature among others, the potential for causing drastic changes in the structure of both aquatic and terrestrial communities. This fact is the more important when one considers that the evolutionary process that has produced the communities themselves has also developed in microbial populations the capacity for controlling the amount of nitrogen, and in certain instances other elements, available to themselves, higher plants and other organisms within the community. Such instances of mutualism are common in nature but they are especially common in the nitrogen cycle where extraordinarily complex systems have evolved to make the molecular nitrogen that forms about 79% of our atmosphere available in various forms suitable for incorporation into life.

At this point in this book every reader has gleaned an acute appreciation for the complexities of living systems. The nitrogen cycle is one of the special wonders of life on earth, complex almost beyond imagination and central to the function of all life. Despite the abundance of nitrogen in the atmosphere and in water as dissolved nitrogen gas, it is not freely available for support of life without a series of chemical transformations to forms that are usable in biotic metabolism.

The transformations of nitrogen that result in its availability for living systems are only a part of the array of biotic transformations that involve nitrogen, including the release of molecular nitrogen back into the atmosphere. We cannot do more here than offer a broad outline of the general cycle and discuss its significance in aquatic systems. The common forms in which nitrogen occurs in nature are listed in Table 11.2. Gaseous molecular nitrogen or N_2 is by far the most common form. The gaseous nitrogen is transformed into biotically active nitrogen by two separate general processes: through the ionizations produced in air by any of several mechanisms including high temperatures

Table 11.2. The forms of nitrogen that are important in the world cycle. Nitrogen occurs in many additional organic forms in nature, but normally enters higher plants as either the nitrate or the ammonium ions.

	Compounds	Solubility
N_2	molecular nitrogen	slightly
NH_3/NH_4^+	ammonia/ammonium ion	readily
N_2O	nitrous oxide	slightly
NO	nitric oxide	slightly
NO_2	nitrogen dioxide	readily
HNO_2/NO_2^-	nitrous acid/nitrite	readily
HNO_3/NO_3^-	nitric acid/nitrate	readily

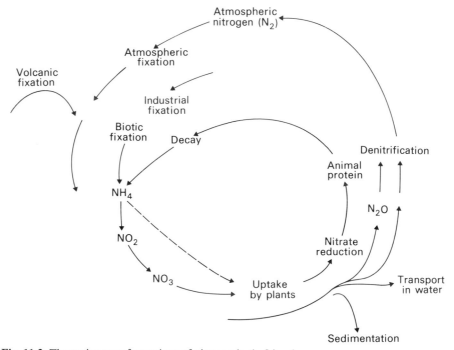

Fig. 11.3. The major transformations of nitrogen in the biosphere.

associated with electrical sparks such as lightning and the operation of internal combustion engines, and by microbial activity. A third process becomes more important annually: the industrial fixation of nitrogen for use in fertilizers. But microbial fixation remains the largest source of nitrogen in support of life.

The complexity of the circulation of nitrogen in the biosphere is indicated in Figs. 11.3 and 11.4, both very much simplied diagrams of the nitrogen cycle. What should be clear from a review of these figures is that the major components of the cycle are microbial transformations that both enable the uptake of nitrogen by higher plants and animals, and mediate under certain circumstances the return of molecular nitrogen to the atmosphere.

The role of nitrogen in the oceans is still more complicated. The major source of nitrogen in support of the oceanic biota appears to be by fixation, although

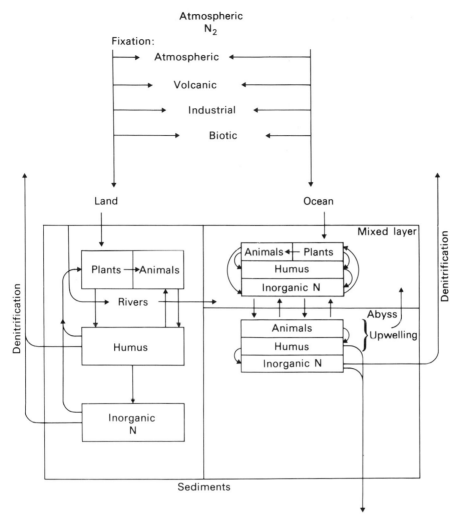

Fig. 11.4. The circulation of nitrogen in the biosphere.

there is obviously a contribution of fixed nitrogen from atmospheric sources and from runoff from the land (Fig. 11.3). There is a constant transfer of mixed nitrogen out of the biotically active zone of the mixed layer into the abyssal waters. This nitrogen is ultimately mineralized through the decomposition of organic matter and may be resident for a period of hundreds of years in the deeper waters of the abyss before being sedimented or transferred again into the mixed layer through upwelling. The upwelling also brings phosphorus to the surface and enables a vigorous new round of carbon fixation in certain segments of the ocean well known for this phenomenon. The most famous of these areas is on the west coast of South America where currents and the configuration of the coast combine to bring nutrient-laden deeper waters to the surface. It is this area that supports for long periods one of the largest fisheries in the world, the anchovy fishery that has yielded as much as 12 million tons of fish annually in a world fishery of 60–70 million tons (see Chapters 7 & 8).

The magnitudes of the fluxes of nitrogen through the biosphere are the subject of continuing research. The best indications at the moment are that the atmosphere contains about 4×10^{21} g (4×10^{15} metric tons) from which a combination of processes results in an estimated uptake of fixed nitrogen by the biota of 4×10^{15} g (4×10^9 metric tons) annually. The processes are shown diagrammatically in Fig. 11.4. While the magnitudes of the fluxes are too uncertain to be set forth here in detail, the important question is the ratio of the man-caused flux to that of the natural flux; how large an effect is man having at present on the nitrogen cycle? Is there a basis for concern that human activities may result in significant enrichment of the biosphere in fixed nitrogen?

The answer is difficult to arrive at, at least with precision. The difficulties arise from the awkwardness of measurements and the necessity for relying on small samples. Present estimates suggest that the man-induced flux, including industrial fixation and fixation through combustion, may be as much as 30% or more of the natural flux. Such a change is substantial and carries with it the information that we can expect to find altered nutrient ratios in forest soils as well as in most water bodies. The implications are that the normal biotic structure of the communities will be altered in the direction of eutrophication and pollution. Such a change is extremely important as a worldwide change, another clear indication that human activities are a major factor affecting the biosphere globally.

11.4 The sulphur cycle and aquatic systems

The recent discovery off the Galapagos Islands of a community of animals living around submarine volcanic vents far below the maximum penetration of light, reminds biologists that H_2S and other reduced sulphur compounds can serve as a source of energy in support of complex communities quite independent of light. The source of energy here is the energy of the earth itself, residual from its formation some 4.5 billion years ago. This complex sulphur-based community is an anomaly in the current general scheme of the biosphere but a reminder that the evolution of life may have had a dark beginning built around hydrogen and sulphur.

The metabolism of sulphur compounds is almost infinitely complex and by no means limited to submarine fumaroles. Microbial transformations of sulphur, including the photosynthetic and non-photosynthetic oxidation of reduced sulphur compounds and the reduction of sulphite and sulphate, occur commonly in fresh and salt water and especially in marshes around the world. These transformations of sulphur, together with the man-caused mobilization of sulphur into the biosphere, involve large fluxes of energy and make details of the sulphur cycle one of the three or four major worldwide topics in ecology.

The source of sulphur in the biosphere is the erosion of primary minerals in the crust of the earth. The amounts available from this source are enormous: terrestrial sediments and unconsolidated surface mineral matters including soils are estimated to contain as much as 5000 kg S/m²; the oceans, a major repository for sulphur once it has been mobilized, contains sulphur in the sulphate form with a surface density of about 2500 kg/m².

The atmospheric component of the world cycle is small because the atmosphere does not usually retain sulphur compounds for more than a day or two. This short residence time means that sulphur compounds do not move worldwide, and local sources of sulphur released into air can be expected to have a more localized or regional effect as opposed to a worldwide effect.

Sulphur is mobilized by various means, including the mechanical mixing of sea spray into the lower atmosphere. By far the most important natural mobilization of sulphur occurs in marshes, however, where microbial activity reduces sulphate and releases H_2S and other volatile reduced sulphur compounds in the atmosphere. In the atmosphere these compounds are oxidized, produce sulphuric acid, and contribute significantly to the local acidification of rain. The importance of the mobilization of sulphur compounds in marshes is a subject of current research.

There is a parallel release of reduced sulphur compounds from terrestrial ecosystems, but at a lower rate.

Superimposed on this normal flux of sulphur into the atmosphere is the mobilization of sulphur through human activities. The largest effect is in combustion of fossil fuel, some of which may contain 3–4% sulphur. There is an additional release to the atmosphere through the smelting of metallic ores that occur as sulphides. A crude approximation of the world cycle of sulphur appears in Fig. 11.5. The important point in this cycle is that the man-caused flux is large by comparison with the natural fluxes. The result is the acidification of rain over the northeastern U.S.A. and northern-western Europe, with attendant effects on lakes, streams, the growth of forests and on agriculture.

The effects of the acidification of rain due to the mobilization of oxides of sulphur and nitrogen are far-reaching, they have become most conspicuous in lakes of the Adirondack Mountains of New York and Scandinavia where the lakes themselves have been acidified to the point where fish populations do not reproduce. Such changes are substantially irreversible and constitute a serious step in the general process of impoverishment. Again a detailed consideration of a natural cycle of a biotically important element shows that human influences are extremely important and growing.

11.5   Toxins and world cycles

The clearest advances in our understanding of world cycles have come from the use of radioactive debris from the bomb tests of the 1950s and early 1960s and from experience with persistent chlorinated hydrocarbons, including the insecticide DDT and the commercial solvents and insulators made with poly-chlorinated biphenyls (PCBs). From the bomb tests which injected large quantities of radioactive debris into the atmosphere we learned details of the circulation of the atmosphere including the fact that air moves around the world in the middle latitudes in a matter of days to weeks and that particulate matter is removed from the air most effectively by precipitation. We learned the residence times of particles of different sizes and the fact that there are systematic exchanges between the upper atmosphere or stratosphere and the lower atmosphere or troposphere. These exchanges occur in the spring in the upper middle latitudes of both hemispheres. We learned further that the transfer of air between the northern and southern hemispheres is limited and that the

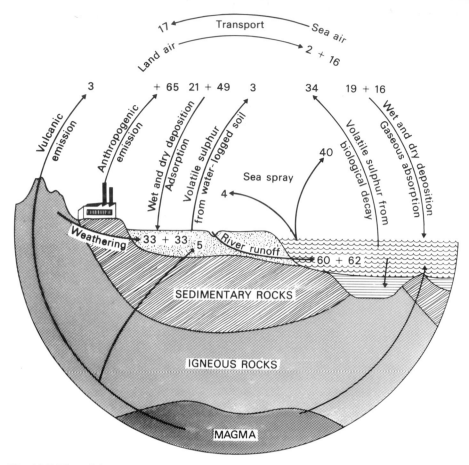

Fig. 11.5. The sulphur cycle in broad outline. Details of the biotic transformations are given in the text. [After Svenson & Söderlein 1976]

transport of radioactivity from the northern hemisphere to the southern hemisphere was also limited, although clearly important.

In addition to these fundamental facts about the physical circulation of the atmosphere we learned that biotic systems have the capacity for absorbing not only nutrient elements that occur in extremely low concentrations in the general environment but also various toxins. Just as the nutrient elements may be concentrated and distributed through complex food webs, so can toxins be distributed differentially throughout the biota. Substances such as DDT and the PCBs, which are extremely soluble in fats and not very soluble in water, accumulate in living systems by an additional mechanism based simply on their differential solubility.

These substances, radioactivity from the bomb tests that continue to be carried out around the world, from the nuclear power industry, and from the reprocessing of military wastes, and other toxins that are sufficiently persistent such as DDT and the PCBs, all have a worldwide distribution. Despite the enormous volume of the oceans, virtually every organism contains detectable

214

residues of man-made substances. The contamination extends to high altitude lakes, glacial ice in the polar regions, and the benthic organisms of the oceanic depths.

These experiences have shown that the intensity of human activities has reached the point where the basic chemistry of the earth is challenged; not only is man mobilizing into the biosphere unusual quantities of major nutrient elements including carbon, nitrogen and sulphur and thereby threatening to upset the ratios that Redfield observed in the sea and considered to have been evolutionarily derived, but man also has the capacity for introducing noxious or even toxic substances into world cycles in quantities that have biotic effects.

Such changes emphasize that science and governments, both national and international, must enter the twenty-first century in a new and evermore intensive alliance if the basic equilibrium of the biosphere on which we all depend is not to be destroyed.

References

ALVERSON D.L. & LARKINS H.A. (1969) Status of knowledge of the Pacific hake resources. *Rep. Calif. coop. ocean. Fish. invest.* **13,** 24–31.

ALVERSON D.L., LONGHURST A.R. & GULLAND J.A. (1970) How much food from the sea? *Science N.Y.* **168,** 503–505.

ANDERSON J.M. & MACFADYEN A. (eds.) (1976) *The Role of Terrestrial and Aquatic Organisms in Decomposition Processes.* Blackwell Scientific Publications, Oxford.

ARX W.S. VON (1962) *Introduction to Physical Oceanography.* Addison-Wesley; Reading, U.S.A.

BARDACH J.E., RYTHER J.H. & MCLARNEY W.O. (1972) *Aquaculture: the Farming and Husbandry of Freshwater and Marine Organisms.* Wiley-Interscience, New York.

BARNES R.S.K. (1974) *Estuarine Biology.* Arnold, London.

BARNES, R.S.K. (ed.) (1977) *The Coastline.* Wiley, London & New York.

BARNES R.S.K. (1980) *Coastal Lagoons: the Natural History of a Neglected Habitat.* Cambridge University Press, Cambridge.

BAYLY I.A.E. & WILLIAMS W.D. (1973) *Inland Waters and their Ecology.* Longman, Camberwell, Australia.

BELL F.W. (1978) *Food from the Sea: the Economics and Politics of Ocean Fisheries.* Westview Press, Boulder, Colorado.

BERRY, R.J. (1977) *Inheritance and Natural History.* Collins, London.

BLUEWEISS L., FOX H., KUDZMA V., NAKASHIMA D., PETERS, R. & SAMS S. (1978) Relationships between body size and some life history parameters. *Oecologia Berl.* **37,** 257–272.

BOJE R. & TOMCZAK M. (eds.) (1978) *Upwelling Ecosystems.* Springer-Verlag, Berlin.

BRAMWELL M. (ed.) (1977) *The Mitchell Beazley Atlas of the Oceans.* Mitchell Beazley, London.

BRANCH G.M. (1975) Mechanisms reducing intraspecific competition in *Patella* spp.: migration, differentiation and territorial behaviour. *J. anim. Ecol.* **44,** 575–600.

BRINKHURST R.O. (1974) *The Benthos of Lakes.* Macmillan, London.

BROCKSEN R.W., DAVIS G.E. & WARREN C.E. (1970) Analysis of trophic processes on the basis of density-dependent functions. In: J.H. Steele (ed.), *Marine Food Chains,* pp. 468–498. Oliver & Boyd, Edinburgh.

BROOKS J.L. & DODSON S.I. (1965) Predation, body size and composition of plankton. *Science N.Y.* **150,** 28–35.

BRYLINSKY M. & MANN K.H. (1973) An analysis of factors governing productivity in lakes and reservoirs. *Limnol. Oceanogr.* **18,** 1–14.

CHAPMAN V.J. (ed.) (1977) *Wet Coastal Ecosystems.* Elsevier, Amsterdam.

COLEBROOK J.M., GLOVER R.S. & ROBINSON G.A. (1961) Continuous plankton records: contributions towards a plankton atlas of the north-eastern Atlantic and the North Sea. *Bull. mar. Ecol.* **5,** 67–80.

CRISP D.J. (1976) The role of the pelagic larva. In: P. Spencer Davies (ed.), *Perspectives in Experimental Biology,* Vol. 1, *Zoology,* pp. 145–155. Pergamon, Oxford.

CUSHING D.H. (1971a) A comparison of production in temperate seas and the upwelling areas. *Trans. Roy. Soc. S. Afr.* **40,** 17–33.

CUSHING D.H. (1971b) Upwelling and the production of fish. *Adv. mar. Biol.* **9,** 255–334.

CUSHING D.H. (1975) *The Productivity of the Sea.* Oxford University Press (Oxford Biology Reader No. 78), Oxford.

CUSHING D.H. (1977) *Science and the Fisheries.* Arnold, London.

CUSHING D.H. & WALSH J.J. (eds.) (1976) *The Ecology of the Seas.* Blackwell Scientific Publications, Oxford.

DAYTON P.K. (1975) Experimental evaluation of ecological dominance in a rocky intertidal algal community. *Ecol. Monogr.,* **45,** 137–159.

DE BEER G. (1958) *Embryos and Ancestors,* 3rd Ed. Oxford University Press, Oxford.

DEEVEY E.S. (1941) Limnological studies in Connecticut. VI. The quantity and composition of the bottom fauna of thirty-six Connecticut and New York lakes. *Ecol. Monogr.* **11**, 413–455.

DICKINSON C.H. & PUGH G.J.F. (eds.) (1974) *Biology of Plant Litter Decomposition.* Academic Press, London.

DINGLE H. (1962) The occurrence and ecological significance of color responses in some marine Crustacea. *Amer. Nat.* **96**, 151–159.

DODSON S.I. (1974a) Zooplankton competition and predation: an experimental test of the size-efficiency hypothesis. *Ecology* **55**, 605–613.

DODSON S.I. (1974b) Adaptive change in plankton morphology in response to size-selective predation: a new hypothesis of cyclomorphosis. *Limnol. Oceanogr.* **19**, 721–729.

EDMONDSON W.T. (1970) Phosphorus, nitrogen and algae in Lake Washington after diversion of sewage. *Science N.Y.* **169**, 690–691.

EDMONDSON W.T. & WINBERG G.G. (1971) *A Manual of Methods for the Assessment of Secondary Productivity in Fresh Waters* (IBP Handbook No. 17). Blackwell Scientific Publications, Oxford.

EL-SAYED S.Z. & McWHINNIE M.A. (1979). Protein of the last frontier. *Oceanus* **22**(1), 13–20.

EVERSON I. (1977) The living resources of the Southern Ocean. Southern Ocean Fisheries Survey Programme. GLO/SO/77/1. F.A.O., Rome.

FAGER E.W. & McGOWAN J.A. (1963) Zooplankton species groups in the North Pacific. *Science N.Y.* **140**, 453–460.

FENCHEL T. (1975) Character displacement and coexistence in mud snails (Hydrobiidae). *Oecologia, Berl.* **20**, 19–32.

FOGG G.E. (1975a) Biochemical pathways in unicellular plants. In J.P. Cooper (ed.), *Photosynthesis and Productivity in Different Environments*, pp. 437–457. Cambridge University Press, Cambridge.

FOGG G.E. (1975b) Primary productivity. In: J.P. Riley & G. Skirrow (eds.), *Chemical Oceanography* 2nd edn., Vol. 2, pp. 385–453. Academic Press, London.

FOGG G.E. (1975c) *Algal Cultures and Phytoplankton Ecology*, 2nd edn. University of Wisconsin Press, Wisconsin.

F.A.O. (1966) *Yearbook of Fishery Statistics*, Vol. 20. F.A.O., Rome.

F.A.O. (1968) The state of world fisheries. *World Food Problems*, No. 7. F.A.O., Rome.

F.A.O. (1972) *Atlas of the Living Resources of the Seas*, 3rd edn. F.A.O., Rome.

F.A.O. (1975) *Yearbook of Fishery Statistics*, Vol. 38. F.A.O., Rome.

F.A.O. (1977) *Yearbook of Fishery Statistics*, Vol. 42. F.A.O., Rome.

F.A.O. (1978) *Yearbook of Fishery Statistics*, Vol. 44. F.A.O., Rome.

GOLTERMAN H.L. (1975) *Physiological Limnology.* Elsevier, Amsterdam.

GOODE R.E., WHIGHAM D.F. & SIMPSON R.L. (1978) *Freshwater Wetlands.* Academic Press, New York.

GOULD S.J. (1977) *Ontogeny and Phylogeny.* Belknap, Cambridge, Mass.

GULLAND J.A. (1971) *The Fish Resources of the Ocean.* Fishing News (Books) Ltd., West Byfleet, Surrey.

GULLAND J.A. (1974) *The Management of Marine Fisheries.* University of Washington Press, Seattle.

HAIRSTON N.G., SMITH F.E. & SLOBODKIN L.B. (1960) Community structure, population control and competition. *Amer. Nat.* **94**, 421–425.

HARDEN-JONES F.R. (1968) *Fish Migration.* Arnold, London.

HARDEN-JONES F.R., ARNOLD G.P., GREER-WALKER M.G. & SCHOLES P. (1979) Selective tidal stream transport and the migration of plaice (*Pleuronectes platessa* L.) in the southern North Sea. *J. Cons. int. Explor. Mer.* **38**, 331–337.

HARDY A.C. (1962) *The Open Sea. Part 1: The World of Plankton.* Collins, London.

HARGRAVE B.T. (1973) Coupling carbon flow through some pelagic and benthic communities. *J. fish. res. Bd Canada* **30**, 1317–1326.

HARPER J.L. (1978) *Population Biology of Plants.* Academic Press, London.

HOBBIE J.E. & RUBLEE P. (1975) Bacterial production in an arctic pond. *Verh. Int. verein. Limnol.* **19**, 466–471.

HUNT H.W., COLE C.V., KLEIN D.A. & COLEMAN D.C. (1977) A simulation model for the effect of predation on bacteria in continuous culture. *Microb. Ecol.* **3**, 259–278.

HUTCHINSON G.E. (1967) *A Treatise on Limnology.* Vol. 2: *Introduction to Lake Biology and the Limnoplankton.* Wiley, New York & London.

HUTCHINSON G.E. (1975) *A Treatise on Limnology.* Vol. 3: *Limnological Botany.* Wiley, New York & London.

HYNES H.B.N. (1970) *The Ecology of Running Waters.* Liverpool University Press, Liverpool.

IDYLL C.P. (1973) The anchovy crisis. *Scient. Amer.* **228**(6), 22–29.

IKEDA T. (1970) Relationship between respiration rate and body size in marine plankton animals as a function of the temperature of habitat. *Bull. Fac. Fish. Hokkaido Univ.* **21,** 91–112.

IKEDA T. (1977) Feeding rates of planktonic copepods from a tropical sea. *J. exp. mar. Biol. Ecol.* **29,** 263–277.

JOHANNES R.E. (1978) Reproductive strategies of coastal marine fishes in the tropics. *Env. Biol. Fish.* **3,** 65–84.

JOHNSON M.G. & BRINKHURST R.O. (1971) Production of benthic macroinvertebrates of the Bay of Quinte and Lake Ontario. *J. fish. res. Bd Canada* **28,** 1699–1714.

JÓNASSON P.M. & KRISTIANSEN J. (1967) Primary and secondary production in Lake Esrom. Growth of *Chironomus anthracinus* in relation to seasonal cycles of phytoplankton and dissolved oxygen. *Int. Rev. Ges. Hydrobiol. Hydrogr.* **52,** 163–217.

JØRGENSEN C.B. (1966) *Biology of Suspension Feeding.* Pergamon, Oxford.

KASAHARA S. (1978) Descriptions of offshore squid angling in the Sea of Japan, with special reference to the distribution of the common squid (*Todarodes pacificus* Steenstrup); and on the techniques for forecasting fishing conditions. *Bull. Jap. Sea Reg. Fish. Lab.* **29,** 179–199.

KEELING, C.D. & BACASTOW R.B. (1977) Impact of industrial gases on climate. In: *Studies in Geophysics, Energy and Climate*, pp. 72–95. National Academy of Sciences, Washington D.C.

KERFOOT W.C. (1977) Competition in cladoceran communities: the cost of evolving defenses against copepod predation. *Ecology* **58,** 303–313.

KREBS J. & DAVIES N.B. (eds.) (1978) *Behavioural Ecology: An Evolutionary Approach.* Blackwell Scientific Publications, Oxford.

KUSNETSOV S.I. & ROMANENKO W.I. (1966) Produktion der Biomassee heterotrophee Bakterien und die Geschwindigkeit ihr Vermehrung in Ryblinsk Stausee. *Verh. Int. verein. Limnol.* **16,** 1493–1500.

LANE P.A. (1975) The dynamics of aquatic systems: a comparative study of the structure of four zooplankton communities. *Ecol. Monogr.* **45,** 307–336.

LIKENS G.E. (1975) Primary production of inland aquatic systems. In: H. Lieth & R.H. Whittaker (eds.), *Primary Productivity of the Biosphere*, pp. 185–202. Springer-Verlag, Berlin.

LOCKYER C. (1976) Body weights of some species of large whales. *J. Cons. int. Explor. Mer.* **36,** 259–273.

LONGHURST A.R. (1976) Vertical migration. In: D.H. Cushing & J.J. Walsh (eds.), *The Ecology of the Seas*, pp. 116–140. Blackwell Scientific Publications, Oxford.

LOPEZ G.R. & LEVINTON J.S. (1978) The availability of microorganisms attached to sediment particles as food for *Hydrobia ventrosa* Montagu (Gastropoda: Prosobranchia). *Oecologia, Berl.* **32,** 263–275.

MALY E.J. (1969) A laboratory study of the interaction between the predatory rotifer *Asplanchnia* and *Paramecium*. *Ecology* **50,** 59–73.

MANN K.H. (1969) The dynamics of aquatic ecosystems. *Adv. ecol. Res.* **6,** 1–81.

MANN K.H. (1975) Patterns of energy flow. In: B.A. Whitton (ed.), *River Ecology*, pp. 248–263. Blackwell Scientific Publications, Oxford.

MARGALEF R. (1968) *Perspectives in Ecological Theory.* Chicago University Press, Chicago.

MARGALEF R. (1978) Life-forms of phytoplankton as survival alternatives in an unstable environment. *Oceanologica Acta* **1,** 493–509.

MARSHALL N.B. (1954) *Aspects of Deep Sea Biology.* Hutchinson, London.

MARSHALL N.B. (1963) Diversity, distribution and speciation of deep-sea fishes. In: J.P. Harding & N. Tebble (eds.), *Speciation in the Sea*, pp. 181–196. Systematics Association, London.

MAY R.M., BEDDINGTON J.R., CLARK C.W., HOLT S.J. & LAWS R.M. (1979) Management of multispecies fisheries. *Science N.Y.* **205,** 267–277.

MAYNARD SMITH J. (1978) *The Evolution of Sex.* Cambridge University Press, Cambridge.

MCLAREN I.A. (1974) Demographic strategy of vertical migration by a marine copepod. *Amer. Nat.* **108,** 91–102.

MELCHIORRI-SANTOLINI U. & HOPTON J.W. (eds.) (1972) *Detritus and its Role in Aquatic Ecosystems.* Mem. Ist. Ital. Idrobiol. Vol. 29, Supplement.

MENGE B.A. & SUTHERLAND J.P. (1976) Species diversity gradients: synthesis of the roles of predation, competition and temporal heterogeneity. *Amer. Nat.* **110,** 351–369.

MENZIES R.J., GEORGE R.Y. & ROWE G.T. (1973) *Abyssal Environment and Ecology of the World Ocean.* Wiley, New York.

MILLER R.J., MANN K.H. & SCARRATT D.J. (1971) Production potential of a seaweed-lobster community in eastern Canada. *J. fish. res. Bd Canada* **28,** 1733–1738.

218

Mori S. & Yamamoto G. (eds.) (1975) *Productivity of Communities in Japanese Inland Waters.* Tokyo. University of Tokyo Press.

Munk W. (1955) The circulation of the oceans. *Scient. Amer.* **193**(3), 96–104.

Murdoch W.W. (1969) Switching in general predators: Experiments on predator specificity and stability of prey populations. *Ecol. Monogr.* **39**, 335–354.

Newell G.E. & Newell R.C. (1963) *Marine Plankton.* Hutchinson, London.

O'Connell C.P. & Zweifel J.R. (1972) A laboratory study of particulate and filter feeding of the Pacific mackerel *Scomber japonicus. Fish. Bull.* **70**, 973–981.

Odum E.P. (1969) The strategy of ecosystem development. *Science N.Y.* **164**, 262–270.

Odum E.P. & de la Cruz A.A. (1967) Particulate organic detritus in a Georgia salt-marsh ecosystem. In: G.H. Lauff (ed.), *Estuaries*, pp. 383–388. American Association for the Advancement of Science, Washington D.C.

Parsons T.R. (1976) The structure of life in the sea. In: D.H. Cushing & J.J. Walsh (eds.), *The Ecology of the Seas*, pp. 81–97. Blackwell Scientific Publications, Oxford.

Parsons T.R., Takahashi M. & Hargrave B. (1977) *Biological Oceanographic Processes* 2nd edn. Pergamon, Oxford & New York.

Patten B.C. (ed.) (1971) *Systems Analysis and Simulation in Ecology.* Vol. 1. Academic Press, New York.

Patten B.C. (ed.) (1972) *Ibid.* Vol. 2. Academic Press, New York.

Patten B.C. (ed.) (1975) *Ibid.* Vol. 3. Academic Press, New York.

Pederson G.L., Welch E.B. & Litt A.H. (1976) Plankton secondary productivity and biomass: their relation to lake trophic state. *Hydrobiol.* **50**, 129–144.

Peer D.L. (1970) Relation between biomass, productivity and loss to predators in a population of the marine benthic polychaete *Pectinaria hypoborea. J. fish. res. Bd Canada* **27**, 2143–2153.

Perkins E.J. (1974) *The Biology of Estuaries and Coastal Waters.* Academic Press, London.

Petipa T.S. (1966) Relationship between growth, energy metabolism and ration in *Acartia clausi* Giesbr. (In Russian.) In: *Fiziologiya Morskikh Zhivotnykh.* Nauka, Moscow.

Pianka E.R. (1978) *Evolutionary Ecology* 2nd ed. Harper & Row, New York.

Pike G.C. (1962) Migration and feeding of the Gray Whale (*Eschrichtius gibbosus*). *J. fish. res. Bd Canada* **19**, 815–838.

Pillay T.V.R. (1976) The state of aquaculture. *The Commercial Fish Farmer* July 1976, 8–11.

Pingree R.D., Pugh P.R., Holligan P.M. & Forster G.R. (1975) Summer phytoplankton blooms and red tides along tidal fronts on the approaches to the English Channel. *Nature Lond.* **258**, 672–677.

Porter K.G. (1973) Selective grazing and differential digestion of algae by zooplankton. *Nature Lond.* **244**, 179–180,

Ramus J., Beale S.I. & Mauzerall D. (1976) Correlation of changes in pigment content with photosynthetic capacity of seaweeds as a function of water depth. *Mar. Biol.* **37**, 231–238.

Reay P.J. (1979) *The Biology of Aquaculture.* Arnold, London.

Redfield A.C. (1958) The biological control of chemical factors in the environment. *Amer. Scient.* **46**, 205–222.

Remane A. & Schlieper C. (1971) *Biology of Brackish Water* 2nd edn. Schweizerbart'sche Verlag, Stuttgart.

Riley G.A. (1970) Particulate organic matter in sea water. *Adv. mar. Biol.* **8**, 1–118.

Rodhe W. (1965) Standard correlations between pelagic photosynthesis and light. *Mem. Ist. Ital. Idrobiol.* **18** Suppl., 365–381.

Rounsefell G.A. (1975) *Ecology, Utilization and Management of Marine Fisheries.* Mosby, St Louis.

Ryther J.H. (1969) Photosynthesis and fish production in the sea. *Science N.Y.* **166**, 72–76.

Sanders H.L. (1968) Marine benthic diversity: a comparative study. *Amer. Nat.* **102**, 243–282.

Sculthorpe C.D. (1967) *The Biology of Aquatic Vascular Plants.* St Martin's Press, New York.

Sheldon R.W., Prakash A. & Sutcliffe W.H. Jr (1972) The size distribution of particles in the ocean. *Limnol. Oceanogr.* **17**, 327–340.

Sieburth J.McN. (1976) Bacteria, substrates and productivity in marine ecosystems. *Ann. Rev. Ecol. Syst.* **7**, 259–285.

Sieburth J. McN., Johnson K.M., Burney C.M. & Lavoie D.M. (1977) Estimation of *in situ* rates of heterotrophy using diurnal changes in dissolved organic matter and growth rates of picoplankton in diffusion culture. *Helgol. wiss. Meeresunters.* **30**, 565–574.

Sorokin Yu.I. (1978) Decomposition of organic matter and nutrient regeneration. In: O. Kinne (ed.), *Marine Ecology*, Vol. 4, pp. 501–616. Wiley, London & New York.

Stearns S.C. (1976) Life-history tactics: a review of the ideas. *Quart. Rev. Biol.* **51**, 3–47.

STEELE J.H. (1974) *The Structure of Marine Ecosystems*. Blackwell Scientific Publications, Oxford.

STEEMAN-NIELSON E. (1975) *Marine Photosyntehsis*. Elsevier, Amsterdam.

STEINHART J.S. & STEINHART C.E. (1974) Energy use in the U.S. food system. *Science N.Y.* **184**, 307–316.

STEWART W.D.P. (ed.) (1974) *Algal Physiology and Biochemistry*. Blackwell Scientific Publications, Oxford.

SVENSSON B.H. & SÖDERLUND R. (eds.) (1976) *Nitrogen, Phosphorus and Sulfur - Global Cycles*. SCOPE Report 7, Ecological Bulletin NFR/22, Stockholm.

SWIFT M.C. (1976) Energetics of vertical migration in *Chaoborus trivittatus* larvae. *Ecology* **57**, 900–914.

THORSON G. (1946) Reproduction and larval development of Danish marine bottom invertebrates. *Meddr. Komm. Danm. fisk.-og Havunders.* (*Plankton*) 4(1), 1–523.

VOLLENWEIDER R.A. (1974) *A Manual of Methods for Measuring Primary Production in Aquatic Environments*, 2nd edn. Blackwell Scientific Publications, Oxford.

WALSBY A.E. (1977) The gas vacuoles of blue-green algae. *Scient. Amer.* **237**, 90–97.

WANGERSKY P.J. (1977) The role of particulate matter in the productivity of surface waters. *Helgol. wiss. Meeresunters.* **30**, 546–564.

WARNER R.R. (1975) The adaptive significance of sequential hermaphroditism in animals. *Amer. Nat.* **109**, 61–81.

WATERS T.F. (1969) The turnover ratio in production ecology of freshwater invertebrates. *Amer. Nat.* **103**, 173–185.

WATSON S.W., NOVITSKY T.J., QUINBY H.L. & VALOIS F.A. (1977) Determination of bacterial number and biomass in the marine environment. *Appl. Env. Microbiol.* **33**, 940–946.

WEIBE P.H. (1970) Small scale spatial distribution in oceanic zooplankton. *Limnol. Oceanogr.* **15**, 205–217.

WEIHS D. (1978) Tidal stream transport as an efficient method for migration. *J. Cons. int. Explor. Mer* **38**, 92–99.

WESTLAKE D.F. (1963) Comparisons of plant productivity. *Biol. Rev.* **38**, 385–425.

WETZEL R.G. (1975) *Limnology*. Saunders, Philadelphia.

WHITTAKER R.H. (1975) *Communities and Ecosystems*, 2nd edn. Macmillan, New York.

WHITTAKER R.H. & LIKENS G.E. (1975) The biosphere and man. In: H. Lieth & R.H. Whittaker (eds.), *Primary Productivity of the Biosphere*, pp. 305–328. Springer-Verlag, New York.

WHITTON B.A. (ed.) (1975) *River Ecology*. Blackwell Scientific Publications, Oxford.

WIBORG K.F. (1976) Fishery and commercial exploitation of *Calanus finmarchicus* in Norway. *J. Cons. int. Explor. Mer* **36**, 251–258.

WILLSON R.B. (1967) Report on biological conditions and water quality in the Red Cedar River as affected by discharges from the Hoover Ball and Bearing Company. Mimeograph; Michigan Department of Natural Resources, U.S.A.

ZENKEVITCH L. (1963) *Biology of the Seas of the U.S.S.R.* (Trans. S. Botcharskaya). Allen & Unwin, London.

Index

225

228